T0291100

Silver Nanoparticles

Silver Nanoparticles

Synthesis, Properties, and Applications

Anna Facibeni

JENNY STANFORD
PUBLISHING

Published by

Jenny Stanford Publishing Pte. Ltd.
101 Thomson Road
#06-01, United Square
Singapore 307591

Email: editorial@jennystanford.com
Web: www.jennystanford.com

British Library Cataloguing-in-Publication Data
A catalogue record for this book is available from the British Library.

Silver Nanoparticles: Synthesis, Properties, and Applications

ISBN 978-981-4968-21-8 (Hardcover)
ISBN 978-1-003-27895-5 (eBook)

Contents

Preface

The use of silver as a purifying agent was recognized nearly six thousand years ago when the Egyptians used it to purify water that had to be stored for long periods of time. The chemical symbol for silver is Ag, which has been derived from its Latin name *argentum*. The metal was named so because of it had brilliant luster. The word *argentum* has its roots in the Greek name for the element *arguros* (*άργυρος*), meaning "bright" or "shining." It is one of the nine elements known since antiquity along with carbon, gold, copper, sulfur, tin, lead, mercury, and iron.

In the 1300s, the Catholic Church specifically chose silver for its cups and trays for the Eucharist to prevent the spread of diseases between priests and practitioners. In 1884, the German physician F. Crade stopped the disease that caused blindness of generations of babies, using a drug whose effectiveness was attributed to silver as an active ingredient. These outcomes are the reason that silver treatments have survived over the years. The power of silver is the reason behind this book.

Nowadays, infections caused by bacteria, parasites, viruses, and fungi are growing even further due to the acquired antimicrobial resistance (characterized by multiple mechanisms) besides intrinsic resistance. Consequently, typical treatments have become ineffective, increasing the risk of spreading. Assistant director general for health security of WHO, Dr Keiji Fukuda, says

> "This is the single greatest challenge in infectious diseases today. All types of microbes—including many viruses and parasites—are becoming resistant to medicines. Of particularly urgent concern is the development of bacteria that are progressively less treatable by available antibiotics. This is happening in all parts of the world, so all countries must do their part to tackle this global threat."

In other words, the issue is urgent.

Textiles can be considered as a major channel for the spread of infectious diseases. To reduce risk from this source, antibacterial agents should be used for textiles. Along with reducing the risk of infection, another incentive for their use if preventing unpleasant odors in textiles. However, frequent washings may damage expensive textiles like silk and cashmere. In some conditions, washing may also be difficult, such as in case of automotive interiors and motorcycle helmets.

Using metals in textiles may be a solution to this problem. In recent years, many books on the use of silver nanoparticles have been published but very few are on their use in textiles. I have carried out studies on silver nanoparticles for several years, not only in terms of the synthesis but also the morphological characterization of the substrate to which they were applied. In this book, I discuss the studies with respect to the various synthesis techniques and properties of silver nanoparticles and their advantages and disadvantages, role in environmental safety, applications, and outcomes.

I hope this book will capture the attention of not only scientists but also students of nanotechnology, textile specialists, and people willing to expand their knowledge of this field.

Anna Facibeni
Autumn 2022

Acknowledgements

I would like to thank the members of the Nanolab team whose collaboration helped me in developing and improving the patent.

A special thank you to the co-authors who trusted me and gave me full autonomy to search for potential customers interested in our technology.

Prof. Alessandra Polissi tried to introduce me to the complex world of bacteria, and I would like to thank her for her patience in explaining me the methods used to check the antibacterial properties of materials.

I am grateful to the team at Transfer Technology Office (TTO) of Politecnico di Milano, particularly Barbara Colombo e Massimo Barbieri, followed me in every step on this adventure.

Thanks to my parents who have always encouraged me, even in moments of despair.

And last but not least, a heartfelt thanks to my, Costanza, Lorenzo, and Danilo, whose love always helps me to go forward and overcome life's obstacles.

Anna Facibeni

Chapter 1

Once Upon a Time Silver

1.1 Through the History

1.1.1 The Importance of Names

If asked, we are all more or less able to give meaning to the word *name* we use it literally or as a concept very often. Webster dictionary can help get a formally correct and concise meaning of such a word: *name* is a "word or phrase by which a person, thing, or class is known."

Quite simple, yet it has implications in our way of life. In fact, every name has its own story whether it is given to a living being or a thing and a name is the result of years and years of transformations and derivations. Etymology studies the origin and the history of linguistic forms. The word etymology itself derives from the Latin word *etymologia* that in turn derives from ancient Greek words *etymons* meaning true, real, and *logos* which means to tell.

Early studies about the origin of names date back to Platone (427–347 BC) and Plutarco (46–127 AD). The first encyclopedic treatise on the origin of names written by Isidore of Seville (560–636 AD) [1] is a work of 20 volumes based on ancient books entitled *Etymologiae*. This text was used throughout the Middle Ages until humanism (15th century).

Silver Nanoparticles: Synthesis, Properties, and Applications
Anna Facibeni
Copyright © 2023 Jenny Stanford Publishing Pte. Ltd.
ISBN 978-981-4968-21-8 (Hardcover), 978-1-003-27895-5 (eBook)
www.jennystanford.com

Languages evolve over time as if they were living things, and for this reason, some words disappear, some new ones appear, and others change. However, some words turned out to keep their meaning over the ages because they are bound to well-defined fields.

The word *chemistry* can be an intriguing example of how a name can remain practically unchanged through the times and cultures while no definite etymology is known.

Although it is widely accepted that the word chemistry derives from the term alchemy, which refers to the ancient practice of treating, purifying, and transmutating materials and metals, the origin of the word alchemy is debated.

According to a number of sources, the word alchemy derives from Arab "al-kimiyà" (الكيمياء). In the Arab language, "al" correspond to the article "the," whereas there is only a hypothesis about the word "kimiyà." For many researchers, kimiyà would come from the Greek language "chemeia" which means fusion. In this case, it could refer to metal fusion. For others, it derives from Egyptian "khem" which means black soil, which is the ancient name of Egypt given after the black color of silt left by the Nile during the flooding. It is the black color that suggests to some scholars the connection of the words "khem" and "kimiyah;" in fact blackening, or *nigredo*, is a very precise moment of alchemical processes, that is, the initial state of chaos or "confused mass."

Other hypotheses try to explain the term as coming from the Chinese "kim-iya," which means "juice to make gold" with clear reference to metal transmutation.

It is generally accepted that the ancient practice of alchemy arose at Alexandria as a child of Greek culture, bearing names chemeia earlier and chumeia later on. However, some sources suggested that alchemy and related nouns arose in China [2] by an aged ascetic who longed for drugs of longevity. He first tried jade, next gold and cinnabar trying to obtain, the ideal drug red like cinnabar and fireproof like gold. What was actually prepared was red colloidal gold, by grinding gold granules in a decoction of a herb of supposed longevity. Such a drug was called Chin-I from Chin which means gold and I which means plant juice. In the local dialect of Fukin, a region of China, Chin-I is pronounced Kim-Iya. From there word went to Arab populations in close contact with China for silk trade, and eventually to Greece by a short step [3].

In chemistry, as it was customary in alchemy, a name of an element is always associated with a symbol, however, we rarely pay attention to their origin. David Ball in 1985 published an interesting paper on the etymology of chemical symbols [4]. Among them, there are some attention-grabbing cases useful to remind how different the symbol is from today's English name, just a few of them are summarized in Table 1.1.

Table 1.1 Few element names

Symbol	Literal origin of the symbol	Current name	Origin of the name that generated the symbol
Au	Aurum	Gold	From Vulgar Latin *orum*, descending from the Latin *aurum*, which comes from the ancient Proto-Greek language ayros χρυσός meaning treasure.
Cu	Cuprum	Copper	From *cuprum*, a distorted name of the Mediterranean island of Cyprus. The origin ascends from the fact that the main sources of Roman copper (*aeramen* in Late Latin, derivative of the Latin *aes*) were the mines of the island of Cyprus, from where the metal came to be known as aes cyprium. Eventually, the word aes became reserved for bronze and cyprium for copper. This word soon was distorted in cyprum, then cuprum.
K	Kalium	Potassium	From neo latinum *kalium*, originated from Arabic *alkali*, which means plant ash. From Antiquity to the Middle Ages potash and soda both compounds were identified as soda, the name potassium appears for the first time in 1648, resulting from Dutch *potaschen* (pot ashes), so called because originally obtained by evaporating the lye of wood ashes in pots.

(Continued)

Table 1.1 *(Continued)*

Symbol	Literal origin of the symbol	Current name	Origin of the name that generated the symbol
Na	Natrium	Sodium	From the Latin word *natrium* a natural salt. *Natrium* comes from the Greek *nitron*, which in turn derived from the Egyptian name of salt "*ntry*," which means pure, divine, adjectives of "*Ntr*" which means god. The substance has been given its name to the ancient mining site, Wadi El Natrun, an almost dry lake in Egypt which contained high amounts of sodium carbonate (Na_2CO_3). *Natrium* was used in the operation of embalming, for its water absorption properties, and was a very important part of religious rituals.
Sb	Stibium	Antimony	From Latin *stibium*, which means fine black powder obtained from the soot of beech used in painting or as a cosmetic. The Latin word in turn derived from Greek στιβι (stibi), the same cosmetic powder (Sb_2S_3) as before.
W	Wolf rahm	Tungsten	From the German *wolf rahm*. It means wolf foam because this metal eats tin as a wolf eats sheep. In the 20th century, some chemists still called it wolfram.

The symbol of silver is another one that is not related to the English name of the element. In fact, Ag, which is the chemical symbol of silver, is originated from the Latin *argentum* and the Greek αργύριον linked to αργός argos meaning "shining, snowy, and white."

Phonetic assonance with Latin and Greek can be found in the European Gaelic airgead as well as when moving away from Europe to get to the India region where one can find *arj-una* as the term for light, white, silver in Sanskrit, although the actual Hindi name of silver is *rajata*.

The Ancient Egyptian Hieroglyphs "*hedj*" represents both white and silver. Silver was very highly prized in Egypt, and rather scarce. It was very popular in pharaonic jewelry, when it was available and was known as "white gold" (*nub hedj*). The white lotus was the symbol of the god Nefertum and statues of him were sometimes made of silver. Silver and gold together represented the moon and sun, respectively.

Words derived from argentum to name the element are still in use in a few languages such as French (argent), Italian (argento), Romanian (argint), etc. On the other hand, the word silver derives from proto-germanic *silubra* from Proto-Indo-European *silubr*.

1.1.2   Silver Diffusion in the Antiquity

Silver, together with copper, iron, gold, and later tin and mercury, was already known to prehistoric populations [5]. Therefore, as it happens to things that have always been with humans since the dawn of history, you cannot say when it was firstly used by humankind.

Among the above metals that can be suitably called "of antiquity," copper, gold, and silver were certainly the oldest and the more important and precious ones in ancient times. Of the three, silver was the last to be discovered probably due to its low presence as metal in its native state.

Due to the ease of manipulation, gloss, resistance to aging and rarity it is easy to think about why ancient civilizations have used silver for luxury and religious related items as well as an exchange mean.

In times as remote as the law code of Menes, the legendary founder of the Egypt I Dynasty (c. 3500 BC), the exchange ratio between silver and gold had already been established. Some thousand years later, it can be read about the silver worth in the *Book of Genesis*. More precisely in Death and Burial of Sara [6], it is written "The years of the life of Sarah was a hundred and twenty... Sarah died in Kiryat-Arba, that is Hebron... and Abraham came to mourn for Sarah... After the dirge, Abraham said to Ephron the Hittite to sell land for the burial of his wife... Abraham agreed with Ephron; and Abraham weighed to Ephron the silver that he had named in the hearing of the Hittites, four hundred *shekels* of silver, current money with the market... ." The *shekel* was an ancient unit of weight (between 10 g and 13 g) used in the Middle East and Mesopotamia. Silver and gold shekel was used as currency.

In the *Odyssey*, the Greek epic poem of Homer dated around the 9th century BC, we can see in various books the description of silver objects like dogs, jambs, and basins.

Other known pure substances in antiquity were metals such as iron, tin, lead, and mercury and elements such as carbon and sulfur. Actually, in past centuries there has been confusion between the various metals, alloys, and compounds very often.

The Dominican friar Vincent de Beauvais, an encyclopedist of the Middle Ages, in his *Speculum Naturale*, which was published toward the middle of the 12th century, attributed to Razes, a renowned Arabian chronicler of the alchemy of the 10th century, a statement that "copper is potentially silver." In practice, according to what de Beauvais ascribed to Razes about copper, "anyone who can eliminate red color will bring it to the state of silver, for it is copper in outward appearance, but in its inmost nature is silver." Furthermore, the encyclopedist said that "pure white mercury, fixed by virtue of white non-corrosive sulfur, engenders in mines a matter which fusion changes into silver" [7].

In China, silver-based artifacts became relatively abundant in the Warring States period (441–221 BC). Silver became a symbol of social status much later than other regions of the world perhaps because of the scarcity of its deposits [8]. Its use became more frequent during the Tang and Song dynasties (7th–13th centuries), probably due to contacts with cultures in Central Asia and the Mediterranean area. In this period the number of mines registered increased and areas like Lingnan, Hunan, and Jiangxi became the major production centers.

In 1513 the Portuguese explorer Jorge Álvares landed on the coasts of Guandong, a Chinese province located on the southern coast of China. The Guandong was the first Chinese region that had commercial relations with the Europeans (from the 16th century), especially through Canton, until 1848 the only Chinese port open to foreign trade.

In 1557 the territory of Macao, south of Canton, was leased from the Chinese Empire to the Portuguese who, without prejudice to the sovereign rights of the empire, had the authorization to create a base for it. The main center of the territory was then a modest seaside resort that came from the Portuguese named Cidade do Santo Nome de Deus de Macau and that was strengthened as a commercial port and as a center of Jesuit missions. Trade between China and Portugal began almost immediately, for a long time it was conducted unofficially and later also clandestinely, so many products and ideas were not introduced to China on a large scale until much later. One exception is military technology: the European cannon was quickly studied and replicated in China and taken to the Great Wall. In addition to military technology, even crops such as tomatoes and potatoes, imported from the Americas, have had a major impact on the Chinese population.

Thanks to the importation of South American silver, the Ming Dynasty of China had moved to a silver-based economy. This metal was increasingly becoming the preferred currency, but China had very few silver mines. This made Peruvian silver very attractive to Chinese merchants, and European merchants could obtain very good prices for silk and Chinese porcelain thanks to this growing demand. The Portuguese empire and the Dutch East India Company made their fortune with this trade, once the Peruvian mines dried up and the flow of silver ceased to circulate in China, the Ming Dynasty was unable to face the resulting economic upheaval with consequent revolts by the army to which salaries were no longer paid.

Crossing the border with India, we can find a classification of silver according to its origin: khanija ('mineral,' obtained from digging mines), sahaja ('natural,' produced or obtained from the peaks of hills), kṛtrima ('artificial,' obtained from mercurial transformation). Until the beginning of the 20th century knowledge of Indian history dated back to a time before the Buddha Gautama period, that is, before the 5th century BC; later, with the discoveries of Harappa and Mohenjo-Daro, the panorama extended up to 3000 BC. The phase matures (around 2750–1900 BC) of the Indus civilization, which some archaeologists prefer to define 'culture of Harappa' since the earliest artifacts of this civilization were found in the urban settlement of Harappa and because this culture extended beyond the boundaries of the river system of the Indus valley, is characterized by the particular expertise demonstrated in the production of metal objects, especially copper and bronze.

Gold and silver jewels, precious stones, inlays, engravings, metal coins and beads were found at Taxila, located north of Rawalpindi, Pakistan, this town shows traces of a long and intense civilization, continued from 500 BC to AD 400 that is, from the era of the Nanda and the Maurya to the Gupta era. Metallurgy is the complex of treatments that must be performed on minerals after mining from mines up to the preparation of metals and alloys.

In India, silver was not widespread in nature, it is thought that it was imported from Burma and Arakan, the ancient population of Kalinga arrived in the 3rd century BC.

1.1.3   Main Uses of Silver in the Past and the Present

1.1.3.1   Silverware and jewels

Around the 2nd century BC, silver was widely used for table services as a characteristic element of the banquets. It symbolized the social status of the host. From the accounts of Roman Poet Martial (c. 41–103 AD), and Pliny the Elder (23–79 AD) we know that many collectors of the time offered a lot of money to buy the works of famous Greek jewelers. Thanks to the Greeks influence, ancient Roman people learned to love good taste, art, and silver.

Likewise, in coeval Mediterranean and Near East civilizations, silver played an important role in pre-Columbian American societies. In the geographic area corresponding to actual Peru, between 1200 BC and 1375 AD, different civilizations prospered, such as Cupinisque, Chavin, Vius and Frias, Moche and Sican in chronological order. Among them, the Moche (or Mochica, 100–600 AD), contemporary with the Nazca civilization settled further south the coast, was one of the most important early-Andean cultures. They were particularly skilled in pottery and metallurgy applied to noble metals, especially silver. From the artifacts that reached up to the present day, we can deduce that the value that they gave to aesthetics. Even the objects of common use such as jewelry, bowls, vessels, tableware, etc. were remarkable artworks, surprisingly similar to ancient Greek ones [9].

Silver usage was continued by cultures that have succeeded in South America, where it was employed in juxtaposition to gold, being the latter restricted to the elite of the Inca Empire [10].

An illustration of the abundance of silver in the Inca Empire is given by the account of the capture of the last Inca, Atahualpa, in late 1532. The ransom paid for his release was defined as a room measuring 7 m × 5 m that should be filled to half its height of 5 m with gold, and twice over to its entire height with silver. It took 2 months to deliver the loot before Francisco Pizarro had him killed regardless. Myth or truth is hard to establish. Maybe that much of the gold could have been tumbaga, the copper-rich alloy made to look like gold through surface treatment (tumbaga composition was roughly Au 60%, Ag 10%, Cu 30%) [11]. In other words, not all that glittered could have been gold.

For sure, more reliable sources to indicate the huge amount of precious metals that came from America at the time can be tax accounts for Seville, Spain's main port for treasure ships [12]. In a little more than one-half of a century, after New World's discovery, about 180 tons of gold and 17,000 tons of silver were legally imported into Europe through the Castile kingdom.

It is precisely at this time that the legend of El Dorado was born. Short for El Indio Dorado, it was the legendary place where an immense amount of gold and precious stones should have been hidden alongside ancient esoteric knowledge. The search for that place lasted for centuries involving others than the Spaniards, even until the `20s of the last century when the English explorer Percy Harrison Fawcett disappeared in the Mato Grosso rain forest (Brazil). Needless to say, nobody has ever found El Dorado.

Back to the Roman civilization again, the use of silver was not limited to tableware and various silverware, but also the ornament of the great celebratory statues. Inscriptions to preserve the memory of honorable Romans were positioned on statues erected in public places. Those silver plates that had attracted the greed for the precious metal were the first to be torn down when the Roman Empire began its decadent parable.

The techniques of silversmith changed over time. In the 1st century AD, the *overhang* method was extensively used. The initial part of the work was done in "negative," modeling the concavity in the slab to obtain the finished work as a relief in "positive." In the following two centuries, the *overhang* went out of fashion and the lost-wax casting and chisel were broadly used.

The lost-wax method, also called *cire-perdue*, was based on the creation of a wax model, then to generate a clay mold where to cast molten silver. In the same period, it was improved a particular method called *niello*. The technique was to fill engraved parts or lines of a silver surface with a mixture consisting of sulfur, copper, silver, and lead (the niello) to be eventually heated to melt by flame. After cooling and polishing non-treated parts, black niello showed up as artistic motifs and writings inlay in the engraving of metals.

Damaged silverware or stuff just gone out of fashion were recast and modeled again very often, however many of them survived history and astonishing objects created by handicraftsmen of the past can be still admired nowadays.

In 1865 during excavations in Boscoreale (Naples, Italy), on the slopes of Vesuvius, came to light the extraordinary treasure of silver, consisting of 111 pieces, hidden in a burlap sack inside of a tank of a country manor by its owner before the eruption of Vesuvius in 79 AD. It was probably a collection made for the ancient hoard, hidden in the danger of invasions with theft and looting.

The pair of silver cups from the Boscoreale treasure (conserved at the Louvre museum) reproduced in Fig. 1.1[1] testify the high quality of the ancient Rome jewelers manufacts. The decoration represents the frailty and caducity of the human condition, skeletons invite you to enjoy the moment. Among the figures of skeletons, we can read: "Enjoy life while you can, tomorrow is uncertain," "Life is a theater," "Enjoy it while you are alive," "The pleasure is the highest good." The choice of depicting the skeletons has nothing macabre, rather it is a hymn to life, an incitement to enjoy the present moment. At the bottom of a cup, we can read the owner's name: Gavia. In the treasury of Boscoreale, we can find also the mirror of injures, Tiberius cup, Apollo cup, the Stork cup, some jugs, and various cutlery.

Another important treasure is that of Hoxne named after the place where it was discovered two decades ago in Mid Suffolk, England. The Late Roman hoard contains an incredible number of pieces now conserved at the British Museum. Including a remarkable silver pepper pot in anthropomorphic form. The hollow vessel is designed as a female half-figure soldered to a separate base. The internal turning disc has two large arcs cut out for the filling holes

---

[1]Figure reprinted with permission of Louvre Museum.

and two groups of perforations designed for dispensing pepper or other expensive food flavoring spice at the dining table.

Figure 1.1 Cups of skeletons in silver (Boscoreale, end of 1st century BC—first half of 1st century AD), h. 10.40 cm × d. 10.40 cm. Louvre Museum Aile Sully, First floor Room Henri II, Room 33, central window 4.

In Berthouville, Eure Department of Normandy, Northern France, accidentally discovered treasure in 1830, turned out to contain the spectacular hoard of gilt-silver statuettes and vessels originally dedicated to the Roman god Mercury. Following four years of meticulous conservation and research at the Getty Villa in Los Angeles, California, the subsequent exhibition of Roman ancient luxury (November 19, 2014–August 17, 2015) allowed visitors to gain new insights and to recognize the full worth and splendor of ancient art, technology, religion, and cultural interaction in Late Roman Gaul (3rd century AD).

One last example worth citing in this non-exhaustive list is the treasure found in Mildenhall, Suffolk, England. Among the beautiful tableware pieces, there was a large concave silver platter (8 kg × d. 60 cm) finely decorated with 135 beads equally distributed on the rim and the entire upper surface decorated in raised relief by chasing and incised lines. The platter, permanently exhibited at the British Museum, is a masterpiece that witnesses once more how much fond of silver were the Romans and the role they played in spreading the appreciation for silver. They used to collect silver from all over the empire as demonstrated by stamped marks on some manufacts of Hoxne Hoard: Trier, Arles, and Lyon (actual France); Aquileia, Milan, Ravenna, and Rome (actual Italy); Siscia (actual Croatia); Sirmium (actual Serbia); Thessaloniki (actual Greece); Constantinople, Nicomedia, Cyzicus, and Antioch (actual Turkey). This piece of information is as important as the treasure itself.

In all the Far East we find examples of antique silver jewelry.

In China, the use of silver in jewelry was often coupled with jade, a mineral consisting of phyllosilicates with chromium inclusions which give it a green color. Confucius (551–479 BC) said that jade possessed eleven virtues, including benevolence (being sweet and lucid), fidelity (it never irritates the skin), education (there was a ritual in clothing) and sincerity (a flaw in jade never hides). Confucian culture preached that a man had to define his ways and conduct in accordance with the virtues of jade.

Diadems, earrings, rings, hair clips, and bracelets have been found in ancient Chinese tombs. In 1968 two tombs were found in modern Mancheng County, in Hebei province. They belong to the prince Liu Sheng (d. 113 BC), son of Emperor Jing Di, and Dou Wan wife of Liu Sheng. Silver and gold needles for acupuncture and burial objects have been discovered in this tomb. Liu Sheng became king of Zhongshan, and he ruled from 154 BC to 113 BC.

In the National Museum of China, in Beijing, we can find many manufacts in gold, silver, and precious stones, among the masterpiece a gold and silver inlay cloud-patterned rhinoceros vessel dating back to the Western Han Dynasty (206 BC–8 AD).

In India, the first jewels date back more than 5,000 years; different in style depending on the region of origin, created for each part of the body. In the region of Himachal Pradesh (northern India), we find the oldest jewels; hair ornaments, long earrings, and large nose rings, as well as elaborate chokers. Along with metals such as silver and copper, tiger nails and teeth were used. The jewelry represented true art like sculpture and painting.

From India, it was easy to go to Tibet, where ancient teapots in embroidered silver have been found.

In Thailand, the votive vases were made of chiselled silver, often in the form of lotus leaves and flowers.

In Myanmar, the use of silver was only allowed to the royals.

In Vietnam, goldwork and silverwork of the Cham culture are preserved from the 10th century. It is exemplified by a crown and heavy jewelry made for a life-size statue found in the ruin of a temple at Mison.

Moving to the American continent more specifically in pre-Columbian American civilization, gold, silver, and copper were the principal metals that were worked, with tin, lead, and platinum used less frequently. When the Spaniards arrived in the New World in the 16th century, they found a wide range of well-developed technical skills in fine metalwork in Mexico, Costa Rica, Panama, and the Andean region.

They knew how to press sheet gold over or into carved molds to make a series of identical forms or sheathing wood, bone, resin, and shell ornaments with gold foil. They decorated with metal, jade, rock crystal, turquoise, and other stones; joining by clinching, stapling, and soldering; possibly drawing gold wire (in Ecuador and western Mexico); casting by the lost-wax method of solid and hollow ornaments, often with false filigree or false granulation decoration; wash gilding; and coloring alloys containing gold by "pickling" in plant acids.

There was some regional specialization: hammer work in "raising" a vessel from a flat disk of sheet gold or silver reached its apogee in Peru and lost-wax casting was highly developed in Colombia, Panama, Costa Rica, and Mexico. Miniature, hollow lost-wax castings of the goldsmiths in Mexico have never been surpassed in delicacy, realism, and precision; and some solid-cast frogs from Panama are so tiny and fine that they must be viewed through a magnifying glass to be appreciated.

In Mexico bimetallic objects of gold and silver were made by two-stage casting; the gold part was cast first and the silver, which has a lower melting point, was then "cast on" to the gold in a separate operation (A famous example is the pectoral of Teotitlán del Camino in the National Museum of Anthropology in Mexico City.). A silver llama in the American Museum of Natural History in New York City indicates that the Peruvian smiths had taken the first step toward cloisonné, enamel decoration technique. It consists in creating, through thin strips of gold, silver or copper, welded to the base metal, compartments (cloisons) to be filled with vitrifiable material.

After vitrification, the surface is equalized and polished and the metal strips appear as lines according to the pre-arranged pattern. The cloisons are filled with cinnabar instead of enamel.

1.1.3.2 The coins around the world

Silver, gold, and copper were also known as *coinage metals*. This classification is due to their use as the main components of the coins from the earliest times to the present, in almost all countries.

The oldest coins found until now date back to an era that is located between the second half of the 7th and early 6th century BC.

The most ancient pieces of evidence came from votive deposits found in a temple dedicated to the goddess Artemis, in the Ionian city of Ephesus, in Asia Minor, actual Turkey. Some of the many small globes of *electrum*, a naturally occurring alloy of gold and silver with trace amounts of copper, were marked in different ways with symbols like lotus flower or animals such as seal, lion, deer, goat, rooster, etc. In addition, one or more geometric punchings were present on the surface to control the core metal goodness inside [13].

The use of marked metal coins was particularly widespread in ancient Greece. Coins soon became distinctive of the different *polis*, the characteristic city-state of Hellenic civilization, with a preference for silver over gold. The coin had the function of being a value unit for goods to be exchanged; in this way, all existing assets could be measured and easily compared in the same units. Besides, it was a payment instrument; each asset could be measured and exchanged with the equivalent value in units of money, the same number of units needed to buy it. Finally, the currency coins could be stored to be used in future transactions.

A marking method technically similar to that of coins was later developed to indicate the quality of metals and metals handmade articles. Such a trademark system was firstly applied to indicate pure silver ingots by officially marking them with a *Tyche*, the Greek goddess of fortune and prosperity. The use of trademarks on objects was progressively disused, those applied after 4th century AD were rare and difficult to interpret.

The Romans had set their own monetary system only during 3rd century BC, well after the birth of their civilization, unifying the bronze experience of Italian origin and that based on precious metals of Greeks. The portrait of the emperor on coins became distinctive of the Roman Empire while coins were made of different materials brass, copper, bronze, silver, and gold, as reported in Table 1.2.

Table 1.2   Roman coins at the time of Julius Caesar

Coin	Make	Size approx.	Worth
As	Copper (bronze earlier)	10 g × d. 30 mm	–
Sestertius	Brass (silver earlier)	25 g × d. 35 mm	2.5 As
Denarius	Silver	4 g × d. 20 mm	4 Sestertius
Aureus	Gold	8 g × d. 20 mm	25 Denarius

The silver *denarius* was the prevailing coinage and during the Republic and the very first period of the Roman Empire, it was of great purity, about 95–98% [14], but was not always remained like that. *Denarius* coins were used to pay troops, and an ever-increasing number of soldiers were required to control an expanding empire that stretched over such a great number of countries. Therefore, to satisfy the growing demand, the purity of the silver used for coins and the corresponding weight was diminished year after year. In fact, it was the number of coins leaving the mints that mattered and not the shape, weight or the title in silver. In those times copper has constantly been coupled to silver to give the coins the necessary hardness while maintaining an optimum amount of silver to the minimum. The proportion of silver to copper in the alloy, however, started to fall during the empire of Nero (54–68 AD) to almost zero in the 3rd century AD. Eventually, the right title was later restored by Emperor Diocletian (244–313 AD) in 296 AD when silver in the alloy was set to 700/1000 (monetary reform).

The legacy of the Late Roman Empire, somewhat characterized by gold coins accompanied by silver coins with a subsidiary role because of devaluation, was transmitted by the Byzantine world also to the Arabic, who gradually went conquering former Roman territories.

In Western Europe, Charlemagne (crowned emperor on Christmas Day 799) extended to all territories of his empire the settle based on silver that had been in use in the kingdom of the Franks [15]. Around 790 Charlemagne introduced a new exclusively silver monetary system after the coin disappeared from economic life in the 6th–8th century because it was replaced by barter. At the base of this new monetary system, there was the weight of a pound of silver (equal to about 408 g), from which were coined 240 coins called a denier, weighing 1.7 g.

Denarius was the only currency circulating, while its multiples existed only as a way of counting. Twelve denarii took the name of solidus or penny (solidus was called the Roman gold coin) and two hundred and forty denarii amounted, as we said, to a pound (hence the word "lira"). Reports 1: 20: 240 (1 lira = 20 money, 1 penny = 12 denari, 1 lira = 240 denarii) remained at the base of the European monetary system until the introduction (after the French Revolution) of the decimal subdivision system. Many monetary units (including the Italian one) have kept for centuries the name "lira" and still at the beginning of the 20th century, a twentieth of lira (5 cents) was commonly called "money."

The Carolingian system was also continued by the Empire Germanicus, with the same pattern of relationships, and spread into neighboring areas, such as England, where the introduced silver *penny* corresponded to the German *pfennig*.

During the Middle Ages, in Europe, there were several types of silver coins in use. In particular the Venetian *Grosso*, a small silver coin of approx. 2 g first introduced at the end of the 12th century, influenced other states to the coinage of the same type of currency. In fact, the *Grosso* had no definite value and it could fluctuate in comparison to other coins.

In the British Isles, it was very common to find a silver penny in the medieval period. Another important currency was the pound, around 1158, but the first coins and banknotes were issued unified since 1694, the year of the founding of the Bank of England. The original name of the pound is a pound of sterling silver, designation indicating an amount equal to a pound of silver 925 (92.5% Ag and the remaining Cu).

The medieval period was characterized by a significant devaluation of the currencies due to large social tensions, deficit growing and imbalance in the balance of payments because of the always ongoing wars. At the turning of 15th century silver coinage in Europe was still poor. Moving forward, in 16th century there were introduced the silver *thaler* in Germany and later on the *ruble* in Russia by Peter the Great. The *thaler* new silver coin was soon used all over Europe and often imitated being the name adapted to fit different languages. The name crossed the Atlantic Ocean and its name originated the actual United States currency, the *dollar*, whose pronunciation is very similar.

The protagonists of the colonization of North America were the Spaniards, the French, the Dutch, and especially the English, who during the 17th and 18th century came to control most of the regions of the eastern coast progressively marginalizing the great rival powers. In this process—in which the Puritan settlers who fled from the homeland to escape religious persecution (Puritanism) played a very important role—finally represented the 7-year war (1756–1763), which recognized Great Britain's domination over much of the country against France and Spain. after the Seven Years' War a serious conflict broke out between the English settlers and the motherland, destined to increase over the next decade. On December 16, 1773, to protest against the British Crown's imposition of a tax on the importation of tea, some American settlers, led by Samuel Adams, boarded British ships and threw loads of tea into the sea (Boston Tea Party). These together with other episodes led to the American Revolution and July 4, 1776, to the declaration of independence of the thirteen colonies drawn up by T. Jefferson (1743–1826) with the birth of the United States. The war broke out with Great Britain that ended in 1783 with the recognition of the independence and sovereignty of the United States, which could count during the conflict on the support of France, Spain, and the United Provinces.

On April 2, 1792, during the presidency of George Washington, the Coinage Act was issued, which ratified the founding of the United States Mint and regulated the issuance of the United States Mint. of the coinage. Treasury Secretary Alexander Hamilton advised Congress to use the dollar as a U.S. monetary unit, following the standard of Spanish dollars and using the decimal system in place of the English duo-decimal system. The deed proclaimed the silver dollar as the basic unit of all other values, thus creating a decimal system of the U.S. currency.

Several types of coins were coined including clad dimes and quarters, that were made of an outer layer of 75% Cu and 25% Ni bonded to a central core of pure Cu. The clad half dollars, with an overall silver content of 40% contain an outer layer of 80% Ag and 20% Cu surrounded to an inner core of about 20% Ag and 80% Cu. U.S. Treasury has minted, as of 1970, about 1¼ billion of these half dollars and placed them into circulation. The composition of the one-cent and five-cent pieces, as well as the silver dollar, remain unchanged. On June 24, 1968, the U.S. Government ceased to redeem U.S. silver certificates with silver.

Ancient Chinese coinage includes some of the earliest known coins. These coins, used as early as the Spring and Autumn period (770–476 BC), took the form of imitations of the cowrie shells that were used in ceremonial exchanges. The Spring and Autumn period also saw the introduction of the first metal coins; however, they were not initially round, instead of being either knife-shaped or spade-shaped. Round metal coins with a round, and later with a square hole in the center were first introduced around 350 BC. The beginning of the Qin Dynasty (221–206 BC), the first dynasty to unify China, saw the introduction of a standardized coinage for the whole empire. Subsequent dynasties produced variations on these round coins throughout the imperial period. At first, the distribution of the coinage was limited to use around the capital city district, but by the beginning of the Han Dynasty, coins were widely used for such things as paying taxes, salaries, and fines.

Ancient Chinese coins are markedly different from coins produced in the west. Chinese coins were manufactured by being cast in molds, whereas western coins were typically cut and hammered or, in later times, milled. Chinese coins were usually made from mixtures of metals such as copper, tin, and lead, from bronze, brass, or iron: precious metals like gold and silver were uncommonly used. The ratios and purity of the coin metals varied considerably. Most Chinese coins were produced with a square hole in the middle. This was used to allow collections of coins to be threaded on a square rod so that the rough edges could be filed smooth, and then threaded on strings for ease of handling.

The payment of silver fees during the Tang Dynasty (618–907) can give us an idea of the amount of silver produced in that period. Considering a 20% tax on metal production, an annual production of about 480–600 kg was estimated during the Tang Dynasty, which reached 42–82 metric tons in the Song period. The monetary function of silver will appear during the Song Dynasty (960–1279); in 1271, the Mongols conquered central China by founding the Yuan Dynasty (1279–1367), the paper currency became the dominant payment system and the silver became the money of account for paper notes.

The growing need for metal money and familiarity with credit notes led Chinese traders during the Tang Dynasty (618–907) to accept the idea of using pieces of paper or paper drafts in payments. The lack of coins also affected the dead to pay their passage to the other world was introduced the funeral paper as funeral money.

At the end of the Tang period, Chinese traders deposited their values at their own corporations, receiving in return notes to the bearer, hequan. Given the success of this practice, merchants were asked to deposit their money in the Government Treasury from now on in exchange for official "compensation notes," called Fey-thsian or flying money. Under the Mongolian government, during the Yuan Dynasty, paper money became the only legal currency. During the Ming Dynasty (1368–1644), the issue of banknotes is given to the Ministry of Finance. Liang was the ancient unit of Chinese weight and mass equivalent to 37.50g, with 98–99% purity. The Chinese notion of Liang for silver was based on syce (silver ingots) until a reform in the 1930s to switch Liang to Yuan. It did not exist a universal weight measure for the silver Liang; the weight varied from 35.14 g to 37.50 g according to the region [16].

In the travel accounts of the Venetian traveler Marco Polo the reader becomes familiar with the fascinating world of paper money production [17]. This money has been put into circulation during the Yuan period by the Mongol chief Kublai Khan (1214–1294): "It is in the city of Khanbalik that the Great Khan possesses his Mint (...). In fact, paper money is made there from the sapwood of the mulberry tree, whose leaves feed the silkworm. The sapwood, between the bark and the heart, is extracted, ground and then mixed with glue and compressed into sheets similar to cotton paper sheets, but completely black (...). The method of issue is very formal, as if the substance were pure gold or silver. On each sheet which is to become a note, specially appointed officials write their name and affix their seal. When this work has been done in accordance with the rules, the chief impregnates his seal with pigment and affixes his vermillion mark at the top of the sheet. That makes the note authentic. This paper currency is circulated in every part of the Great Khan's dominions, nor dares any person, at the peril of his life, refuse to accept it in payment."

Whereas the alchemists had struggled vainly for centuries to turn base metals into gold, the Chinese emperors had very simply turned paper into money. Once back home, Marco Polo amply reported about his experiences and adventures in the Chinese Empire but when he talked about paper money he only met skepticism. It will still take some centuries before paper money will be introduced into Europe.

Official coin production was not always centralized but could be spread over many mint locations throughout the country. Aside from officially produced coins, private coining was common during many stages of history. Various steps were taken over time to combat private coining and limit its effects in making it illegal. At other times private coining was tolerated. The coins varied in value throughout history.

Some coins were produced in very large numbers—during the Western Han, an average of 220 million coins a year were produced. Other coins were of limited circulation and are today extremely rare—only six examples of Da Quan Wu Qian, big coins, from the Eastern Wu Dynasty (222–280) are known to exist. Occasionally, large hoards of coins have been uncovered. For example, a hoard was discovered in Jiangsu containing 4,000 Tai Qing Feng Le coins and at Zhangpu in Shaanxi, a sealed jar containing 1,000 Ban Liang coins of various weights and sizes, was discovered.

The first mention of coins in the context of South Asia can be found in the Vedas (c. 1000–1500 BC), a complex of sacred texts from which the oldest religion of the arias populations of India (vedismo) takes its name, from which later Hinduism will develop. However, the archaeological discoveries suggested that the earliest evidence of coin circulation in India can be traced back to 6th–5th century BC. The more ancient coins are known as the famous 'punch-marked' coins. The punch-marked coins were mainly made of silver; they had a rectangular shape and had printed small symbols, such as elephants, sun, and moon. The weight system of the punch-marked coins as well as all the other ancient Indian coins was based on the red and black varieties of seeds of a tree called *Abrus precatorius*. The weight of these seeds was known as *rattis*. The uniformity of the weight system was one of the main reasons behind the long usage of these coins over a vast area of South Asia [18].

The lack of a precise reference to the person issuing the currency was one of the reasons that allowed the Indian currency to survive longer than the monetary traditions of Europe and the Islamic world. In Europe, from Greek times, the use of portraits, religious imagery, heraldic devices and inscriptions has been systematically used to identify the issuing authority responsible to produce coinage and its circulation. In the Islamic world, the inscriptional designs refer to the authority of the Islamic religion of those producing and regulating coin issues and normally also name the mint and date of issue and later the ruler responsible. In contrast to what happened in Europe and in the Islamic world, in India, the identity of the issuer is completely absent from the coin. The most extreme examples are the imitation-Sasanian issues of western India, which continue for about seven centuries without systematic indication of the issuer, either by dynasty, royal name or mint name. The function of the design seems to be simply to identify the coin as current money. The closer the design is to that of earlier coins the better it corresponds to this purpose.

Rupee is the name of the currency, derived from Sanskrit *rūpya* which means wrought silver, coin of silver [19].

1.1.3.3 Photography

When we talk about silver the first thought goes to its use in photography. The word photography is derived from the Greek words φῶς which means light, and γραφή which means writing, then with photography we can draw with the light. The English scientist Sir John Herschel (1792–1871) in 1839 coined the use of the terms photography, positive, and negative to refer to photographic images used for the first time the term photography, the year the invention of the photographic process was made public.

The discovery of light-sensitive materials has been known for many years. In 1727 German chemist Johann Heinrich Schulze discovered the sensitivity of silver nitrate, 50 years after the Swedish chemist Scheele observed the action of light on silver chloride, darkness, and its indissolubility in ammonia.

The first true photograph was taken by Joseph Nicéphore Niépce (1765–1833) in 1826 on Bitumen of Judea[2] (a substance that possesses the property of becoming insoluble in lavender oil after it has been exposed to light). It is known as "View from the Window at Le Gras." The exposure time of 8 hours causes the impression that the buildings are illuminated by the sun both from the right and from the left. In reality, this was not the first photo made by Niépce. In 1816 he obtained his first photographic image (which depicted a corner of his workroom) using a sheet of paper sensitized, probably, with silver chloride. The image, however, could not be fixed permanently, instead of what he did in 1826 with this "photo" at the Bitumen of Judea.

In the UK the English scientist Thomas Wedgwood, son of Josiah Wedgwood, the well-known potter, and his associate, the English scientist Sir Humphrey Davy made images on paper and white leather coated with silver nitrate. They laid leaves and paintings on glass upon the sensitive materials and exposed them to sunlight, which darkened the silver. In an attempt to keep the image, they washed the exposed materials without success. They found that combining the silver solution with sodium chloride produced the more sensitive whitish paste of silver chloride.

The French painter Louis-Jacques-Mandé Daguerre was led to experiment with the photographic process with his passion for art and illusion; he had in fact opened two Dioramas, pictorial exhibitions enriched by effects due to continuous changes of light, in Paris and London. In 1829 he began a collaboration with Joseph Nicéphore Niépce and his son Isidore, in January 1839 the French Academy of Sciences announced the invention of the process of the daguerreotype, a metal plate, covered with a thin leaf of pure silver, sprinkled with iodine. It had to be used within an hour of its preparation, placing it in a dark room with a lens. The exposure time varied from 15 to 30 minutes, the image was then made visible after the development carried out with mercury vapors. It was a positive image, that is, with non-inverted black and white, very fine in details and shadows, a unique piece that cannot be reproduced. The French government acquired the rights for the public use of the daguerreotype and a permanent pension was assigned to Isidore and

---

[2]A mixture, soluble in lavender oil, containing bitumen, oil from linen, clay, and turpentine essence.

Daguerre. The daguerreotype process spread throughout Europe thanks to a manual written by Daguerre himself, which was then translated into different languages and the reduction in exposure times possible thanks to the progress of chemistry, becoming the main method for creating photographs.

In 1834 the English researcher William Henry Fox Talbot (1800–1877) began his experiments with silver chloride, to get to the calotype using silver iodide more sensitive to light. The paper treated with iodine was then sensitized with silver nitrate, acetic acid, and gallic acid. The paper was exposed to moisture, then developed with gallic acid and stabilized with potassium bromide and fixed with sodium thiosulphate. However, the negative had to be retouched with graphite to avoid the transmission of light. The research continued until 1878 when George Eastman developed a preparation of dry gelatin-bromide plates and in 1884, founded the Eastman Dry Plate Co. and the Film Co., he began the production of transparent films. His first devices, baptized Kodak, went on sale in 1888 and quickly conquered the market of photography and cinematography. In 1889 Eastman manufactured, with the collaboration of HM Reichenbach, the transparent film of nitrocellulose of 35 mm width, base, until the advent of digital, of the film industry (also Thomas Alva Edison used it for his cinematography experiments). In 1892 stood the Eastman Kodak Co., the first company that produced folding cameras and film cameras with reduced pitch (16 mm) on a large scale and at popular prices.

Photography is a permanent record of an image formed on a light-sensitive surface. We can distinguish five fundamental phases in the photographic process (Fig. 1.2).

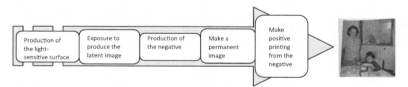

Figure 1.2 Schematic diagram of the photographic process.

The light-sensitive surface is an emulsion of a gelatin silver halide spread on a transparent support. The halide crystals must be small and uniform (less than microns), they are commonly called grains.

The bromide is the most widely used, since AgBr is sensitive to violet and blue radiation they must add sensitizers (usually carbocyanine) which absorb to longer wavelengths, and transfer energy to AgBr.

When, on exposure of the emulsion to light, a photon of energy impinges on a grain of AgBr, a halide ion is excited and loses its electron to the conduction band, through which it passes quickly to the surface of the grain where it can liberate an atom of silver:

$$Br^- + hv \rightarrow Br + e^- \qquad (1.1)$$

These steps are, in principle, reversible but, in practice, are not because Ag is clearly liberated on crystal dislocation, or defect, or at an impurity site such as may be provided by Ag_2S, all of which allow the electron to reduce its energy and so become "trapped." The function of the dye sensitizers is to extend the sensitivity of the emulsion across the whole mechanism for transferring the energy to X^- to excite its electrons. As more photons are incident on the grain, so more electrons migrate and discharge Ag atoms at the same points. The parallel formation of Br atoms leads to the formation of Br_2 which is absorbed by the gelatin.

The development or intensification of the latent image is brought about by the action of a mild reducing agent whose function is to selectively reduce those grains which possess a speck of silver, while leaving unaffected all unexposed grains. To this end, such factors as temperature and concentration must be carefully controlled and the reduction stopped before any unexposed grains are affected. Hydroquinone is a common developer and the reduction is a good example of a catalyzed solid-state reaction. Its mechanism is imperfectly understood but the complete reduction to the metal of a grain starting for a single speck represents a remarkable intensification of the latent image of about 10^{11} times allowing vastly reduced exposure times; this is the very reason for the superiority of silver halides over all other photosensitive materials.

After development, the image on the negative is fixed by dissolving away all remaining silver salts to prevent their further reduction. This requires an appropriate complexing agent and sodium thiosulphate is the usual one since the reaction goes essentially to completion and both products are water-soluble [Eq. (1.2)].

$$AgBr_{(s)} + 2\ S_2O_3^{2-} \rightarrow [Ag(S_2O_3)_2]^{3-} + Br^- \qquad (1.2)$$

A positive print is the reverse of the negative and is obtained by passing light through the negative and repeating the above steps using a printing paper instead of a transparent film.

Bromide absorbs a photon and moves to an activated state. At this stage, the film is dipped in a reducing solution (reducing agents mostly used are organic polyphenols, such as hydroquinone) [20] whose electrochemical potential is slightly smaller than the activated state of AgBr. For this reason, AgBr* will reduce before AgBr, then the latent image is formed.

Now the developed film is dipped in a solution of sodium thiosulphate $Na_2S_2O_3$ (fixing bath) which removes AgBr not reduced.

Finally, the film may be exposed to light. Silver remaining in the baths is recovered by chemical means (e.g., precipitating as silver sulfide Ag_2S) or by electrochemical methods.

These days few companies, such as Kodak, Fujifilm, Ilford, and Ferrania, produce photographic films.[3]

Another important use of photosensitive material is the detection of ionizing radiation.

Dosimetry films although obsolete as a method for assessing exposure to ionizing radiation is still present in some nations. The sensitive material is constituted by two different substrates in cellulose on which two different emulsions sensitivity are deposited, all enclosed in a plastic casing-proof light. This feature ensures the maintenance of information over time without having to destroy partially the image produced by the emulsions high sensitivity using the well-known practice called "stripping," using individual films on which are deposited two emulsions with different sensitivities. The sensitive material is inserted within castings in a polymer ABS (acrylonitrile butadiene styrene) on which are present five areas differently filtered. Only one part is in the air, the others with four filters of different composition and thickness (0.05 mm Cu, 0.3 mm Cu, 1.2 mm Cu, 0.8 mm Pb). The filters are present on both sides of the dosimeter, through the use of badge clips either with the front or back side facing the radiation field.

The application of appropriate calculation algorithms allows estimating the magnitudes Hp (10) and Hp (0.07)[4] [21]. You can also get information about the irradiation mode of the dosimeter and the energy of the incident radiation.

---

[3]There is a very interesting video at www.filmferrania.it

[4]Personal dose equivalent, Hp(*d*), is defined by the ICRP (the International Commission on Radiological Protection) as the dose equivalent in soft tissue at an appropriate depth, *d*, below a specified point on the human body. The specified point is usually given by the position where the individual's dosimeter is worn.

1.1.3.4   Awards and mirrors

It is a piece of metal, usually of a similar shape to that of the coins, melted or coined, and destined to remember a certain person or fact. The medal differs from the currency, to which it is very similar, because it was coined and issued by private individuals (but in some cases also by states) for commemorative purposes and above all, it is not a bargaining chip because it has no face value. The medals look like coins for their shape and appearance, but they are often larger and have more often been merged than jokes. They are not subject to wear, like coins in circulation. Medals were widely distributed in ancient Rome. During the Renaissance period, their use came back to life thanks to the work of Antonio Pisano, called Pisanello (1395–1455), who coined a very famous medal of Byzantine Emperor Giovanni VIII the Palaeologist in 1438, whose wide circulation pushed many Italian lords to emulation [21]. It was the first effigy of a living on a coin since antiquity, a few decades ahead of its monetary emissions. "His medals provided a portable portrait relief of the sitters, reproducible by casting in lead or bronze and small enough to be held in the hand. He placed a profile portrait on the obverse and an allegorical or pictorial scene on the reverse. This formula for the medal has lasted to the present day." The imperial model remained binding, with the effigy always in profile. During the 15th century, the activity of the medalists was limited to the circle of the courts and their most illustrious collaborators. After Pisanello were famous medal makers Matteo de 'Pasti, Cristoforo di Giovanni Matteo Foppa also known as Caradosso, Leone Leoni, up to Benvenuto Cellini, in the 16th century. It represents an honor that is assigned to officers and soldiers decorated with military value, or even to private citizens, and sometimes to institutions or cities and countries, for civil merit or in memory of their participation in events of national importance. It is awarded as a prize to athletes who ranked first in sports competitions. Depending on the importance they are minted in gold, silver, bronze.

We find also Medal of Honor (United States military decoration), Frost Medal (American poetry award), Presidential Medal of

Freedom (American award), Spingarn Medal, Copley Medal (British award), Fields Medal (award), Newbery Medal (literary award), and Caldecott Medal (literature).

The word mirror derives from the Latin *specŭlum*, derived from *specĕre* which means to look, observe. In the mirrors, the surface must be perfectly reflective so as to reproduce a perfect image. The first mirrors made in antiquity were simple sheets of metal, often silver, copper or bronze, perfectly polished.

In the 14th century in Venice mirrors were produced by combining a sheet of polished crystal with sheets of tin and mercury: the thin layers of tin were joined to the glass by a mercury bath and exerting pressure; this process was expensive and complex, making the mirror a luxury product. In the second half of the 19th century, the cost of mirrors has radically decreased thanks to a new production process, the silvering: through a solution of silver nitrate, ammonia, and tartaric acid, to the glass are set very minute particles of silver, which was subsequently covered with shellac.

The chemical process of coating a glass surface with metallic silver was discovered by the German chemist Justus von Liebig in 1835, and this advance inaugurated the modern techniques of mirror making. Today the mirrors consist of a sheet of glass on which a thin layer of silver or aluminum is deposited, fixed to the glass by electrolysis or by sputtering. The metal layer is placed on the opposite side to the reflecting side and is covered with protective paint. In this way, the delicate coating is protected by the glass itself, but there is a second minor reflection caused by the front surface of the glass. This type of mirror reflects about 80% of the incident light. Modern technologies have also reintroduced the production of metal mirrors especially as design objects, or for safety reasons in road mirrors, that is, polished stainless steel to make it look like a traditional glass mirror.

1.1.3.5 Brazing and soldering

The welding is the union of two elements made of the same metal (or similar metals) heated locally at a temperature higher than the melting temperature; if necessary, a "filler material," compatible with the base metals, used to fill the welded joint can be added to the casting bath. There is also a particular type of welding, called brazing, in which the junction between the two metals is realized by melting only the filler metal.

In order for the process to be carried out, there is a need for heat to heat the metal and allow the filler alloy to melt and to penetrate, by capillarity, between the parts to be welded; this heat is mainly given by a flame coming from a torch fuelled by gas (methane, propane, acetylene, etc.) and air or oxygen.

An effect of the heat on the metal is oxidation, which prevents the soldering alloy from flowing, so it is necessary to protect the affected surfaces with a flux, the most used is the borax dissolved in water.

The alloys used for welding are mainly composed of gold, silver, copper, and zinc in various proportions according to use and can be kept, medium or strong according to their fusion point (in the sense that will be more resistant as the temperature necessary for their fusion increases remain, however, always below the metals point of fusion to be welded). These alloys, which are generally shaped of plates or wires, they can be prepared directly in the laboratory or found in the appropriate supplies.

Silver brazing, frequently called "hard soldering" or "silver soldering," is a low-temperature brazing process with rods having melting points ranging from 1145 to 1650°F (618 to 899°C). This is considerably lower than that of the copper alloy brazing filler metals. The strength of a joint made by this process is dependent on a thin film of silver brazing filler metal.

1.1.3.6 Energy, electronics, medicine

Silver has the highest electrical and thermal conductivity of all metals, and it is the most reflective. These physical properties make it a highly valued industrial metal, especially when used in solar cells.

Silver is actually a primary ingredient in photovoltaic cells, and 90% of crystalline silicon photovoltaic cells, which are the most common solar cell, use a silver paste. What happens is that when sunlight hits the silicon cell it generates electrons. The silver used in the cell works as a conductor to collect these electrons in order to form a useful electric current. The silver then transports the electricity out of the cell so it can be used. Further, the conductive nature of silver enhances the reflection of the sunlight to improve the energy that is collected [23].

This increased demand for silver could have a real impact on the solar marketplace in the years to come as solar could push up the price of silver. So, should silver prices surge so that it could have an impact on the production costs of solar panels, which would then impact the economics of the solar industry?

The reflectivity of silver gives it another role in solar energy. It reflects solar energy into collectors that use salts to generate electricity.

Nuclear energy also uses silver. The white metal with boron and indium alloy is often employed in control rods to capture neutrons and slow the rate of fission in nuclear reactors. Inserting the control rods into the nuclear core slows the reaction while removing them speeds it up [24].

Electronics is an important field of use of silver in industry. The thermal and electrical conductivity of silver is superior to other metals it means that cannot easily be replaced by less expensive materials.

Contact materials in relays are very important for their efficiency. The main contact materials normally used for relays with nominal contact ratings between 5 and 50 A are silver/nickel, silver/cadmium oxide, and silver/tin oxide [25].

We can find the use of silver also in ayurvedic medicine; therapeutic remedies date back to the 6th–5th century BC by Charaka, one of the main founders of this alternative medicine. Ayurveda is a word composed of *ayur*, lifespan or longevity, and *veda* which means revealed knowledge. In ayurvedic medicine, the traditional medical lore of Hinduism, *rasa shastra* is a process by which various metals, minerals, and other substances, including mercury, are purified and combined with herbs in an attempt to treat illnesses. *Rasa shastra* [26] is a pharmaceutical branch of the Indian system of medicine that mainly deals with metals, minerals, animal origin product, and toxic herbs and their use in therapeutics. Silver also enjoyed an important place in Ayurveda medications. An example is *rajata bhasma*, an ayurvedic drug used for anti-aging, cardio-vascular diseases, and urinary disorders. It is a combination of metallic silver (52% to 59%), free sulfur (0.675%), ferric oxide (14.33%), calcium (10.769%), silver chloride (0.479%), and traces of sodium, potassium, and aluminum. The normal dosage range given for *rajata bhasma* was 30 mg to 120 mg Classics of alchemy says that samples of rajata which are clear, lustrous (*swachha*), heavy (*guru*), and with metallic sheen (*snigdham*), and which also become bright white on heating or cutting (*dahe chede samaprabham*), without any ridges or furrows (*sphota rahitam*), is genuine, and can be considered acceptable for therapeutic purposes [27].

Silver is used as an antibacterial for centuries. Before the discovery of antibiotics, colloidal silver was considered one of the most powerful remedies against infection, it seemed to be effective against more than 650 different pathogens, including viruses, bacteria, and fungi. It has been used in the past for many diseases such as arthritis, burns, pneumonia, salmonella, lupus, syphilis; its renaissance took place in the seventies in colloidal form, the only one who can be legally used as a supplement, because as the drug was banned in 1975 by the Food and Drug Administration (FDA) for its toxic effects.

Colloidal silver is tasteless and odorless, it was taken orally or put on gauze to treat wounds, burns, warts, and eczema or general skin irritations.

Argyria is a pathology caused by to prolonged assumption of silver colloids or silver ions. This disease is characterized by gray to gray–black staining of the skin and mucous membranes produced by silver deposition.

Traditionally drops of silver nitrate were applied into the eyes of newborns to prevent bacterial eyes infections, now most hospitals use erythromycin ointment as a substitute.

Its use was discontinued in the early 20th century because of the high production costs, and to the advent of penicillin and antibiotics, more economical and effective; the current processes of production of silver have allowed a reduction of the price and have contributed to its return among the remedies available to everyone.

1.1.4 Curiosities

It is stated before that words derived from Latin *argentum* are still in use in some current languages such as in French (*argent*), Italian (*argento*), Albanian (*argjend*), Irish (*airjead*), Romanian (*argint*); as well as those derived from Proto-Indo-European *silubr* are present in English (silver), German (*silber*), Dutch (*zilver*) to name some.

The term for coins in East Europe was dinaro from Latin *denarius*.

In Spanish language, it is used plata for silver; the origin is the medieval Latin *plata* which means plate, piece of metal, perhaps from Greek *platys* which means flat, broad.

A nation takes its name from silver: it is Argentina, and it is associated with the legend of the Sierra de la Plata, common among the first European explorers of the region, both Spanish and Portuguese. When the first Spanish conquistadors discovered the Río de la Plata, they named the estuary Mar Dulce[5] (Sweet Sea). Indigenous people gave gifts of silver to the survivors of a shipwreck led by Juan Díaz de Solís, who was later killed by them. Legend has reached Spain around 1524, and the name was put in print for the first time in a Venetian map of 1536. The silver source was the device by which in 1546 was founded the city of Potosi. The name Argentina was mentioned for the first time in the poem published in 1602 by the Spanish priest Martín del Barco Centenera entitled *La Argentina y conquest of the Río de la Plata*, which describes the region of the Río de la Plata and the founding of the city of Buenos Aires. He then began to be used extensively in the book of 1612, *Historia del descubrimiento, población, y conquista del Río de la Plata by Ruy Díaz de Guzmán*, in which the territory was called Tierra Argentina (Land of Silver). The name is set by the Constitution Argentina, which in the first part of Article 35 explicitly states that: the names adopted after the inauguration of the first national government in 1810 as Provincias Unidas del Río de la Plata (United Provinces of the River Silver), Argentina, Confederation Argentina, and the Argentine Nation are all officially recognized for the name of the government and of this State [28].

Commonly speaking, we can find some frequent sayings containing silver as a noun, some of them are summarized in Table 1.3.

---

[5]Total area 3.2 million square kilometres

Table 1.3   Common sayings comprising the noun "silver"

Silver referring term of common use	Way of saying, notes
Born with a silver spoon in one's mouth	If you say someone was born with a silver spoon in his/her mouth, you mean he/she was born into a wealthy family.
Every cloud has a silver lining Every silver lining has a cloud	Both phrases are referring to bad (clouds) and good situations; John Milton coined the phrase 'silver lining' in *Comus: A Mask Presented at Ludlow Castle*, 1634.
Being a silver surfer	Nothing to do with Marvel's comic character. It refers to an aged person who spends a lot of time using the internet (Cambridge dictionary).
A silver bullet is a complete solution to a large problem	The allusion is to a miraculous fix, otherwise portrayed as 'waving a magic wand.' This figurative use derives from the use of silver bullets in the widespread folk belief that they were the only way of killing werewolves or other supernatural beings.
Silver laugh	Another way of saying clear, softly resonant sound. Sometimes referring to a beautifully clear voice.
Silver anniversary	Twenty-fifth anniversary, usually of the wedding. Back in the Roman Empire husbands crowned their wives with a silver wreath for this recurrence.

Although the two are not chemically related, in ancient times the mercury was considered as a special kind of silver—hence the traditional name of *quicksilver* and its Latin name *hydrargyrium* (liquid silver). The Latin name originated the mercury symbol Hg.

An Italian way to describe a person, especially a child, who is very restless, is saying that he/she is "having living silver," mercury is an element which droplets tend to chase each other. For over a thousand years alchemists considered the metal as a key to the transmutation of base metals to gold and employed it for amalgams both for gilding and for producing imitations of gold and silver.

Although precious metals can be used in their pure form, due to their cost, and above all for their extreme malleability, silver has very often been used in alloys with other metals such as copper, lead, tin, and even gold. The most extensively used metal with silver in alloys is copper, its content depending on the planned use of the alloy. In the case of most valuable objects, a copper concentration of 1% to 4% was used. A higher copper content between 10% and 40% was used to manufacture refined objects of common use such as mirrors and spoons. Alloys containing 20% to 40% of copper were used in welding work for objects of value. By contrast, silver and copper alloys tend to be oxidized and the formation of tarnish superficial films can occur when exposed to air. Copper is firstly oxidized to the corresponding black oxide CuO, long before silver forms the oxide Ag_2O. Similarly, the oxidation by hydrogen sulfide to form the sulfides is easier on copper than on silver. Noteworthy the sulfides CuS, Cu_2S, and Ag_2S can even be formed for exposures to the sub-ppm levels of hydrogen sulfide naturally present in the air. Along with the above oxides and sulfides, the other common product of corrosion visible on the surface of silver-copper alloys is the emerald green malachite ($CuCO_3 \times Cu(OH)_2$), which is in turn, derived by the action of water and carbon dioxide naturally present in the air on copper oxide. All in all, it is the less noble metal of the alloys to be selectively attacked first through a galvanic corrosion mechanism.

Silver and its alloys have been the most widely used materials for currency in history since about 3000 BC, therefore it should not be surprising that throughout the ages it has been tried to remedy the wearing out of coins. A selective corrosion process similar in principle to the above was extensively applied on purpose to renovate used currencies in the times of inflation during Roman Empire, especially between 63 AD and 260 AD. Silvery surfaces with new stamps from worn coins were obtained on a large scale by a three-step process. The material was firstly immersed in a solution of an organic acid, such as acetic acid from vinegar, to dissolve copper and obtain a superficial porous layer of silver. The surface was then polished and finally coined again. The process was legally applied by the empire currency administrators, however, a few years ago it was discovered that an illicit disguise associated with silver coins was carried out in the same period. It was found out that in Rome, by means of forging technology not previously reported, they were able to cover a worthless coin-shaped lump of lead with a bi-layer coating made of copper (inmost) and fine silver. The resulting quality was surprisingly like the result that can be obtained by modern electroplating techniques [29].

In addition to plating with the noble metal, the silvery appearance could be obtained even though the noble metal was not actually present, for example, in the case of arsenical bronze. In fact, techniques to integrate other metals into copper alloys to obtain materials that looked like silver were known since antiquity. These techniques were used to produce original objects, but also for counterfeiting purposes not limited to fake coins.

Examples of alloys with rather low quantitative of noble metal while maintaining an appearance of preciousness was found in Japan as well, in the 18th and 19th centuries.

The two most esteemed Japanese alloys were shakudo and shibuichi where the silver/copper ratio could be varied to achieve the desired polychrome effect. Shibuichi, which means "one-fourth" indicates the standard formulation of one part of silver to three parts copper, however different proportions can be applied to yield colors. The silver percentage can range from 2% to 60% or more, but more often would fall in the 15% to 40% range. These percentages were not randomly used; based on centuries of experience they were calculated for a specific color result. Shibuichi patinated with the usual rokusho irotsuke technique (traditional soft-metal alloy patination) offered a wide range of grays with some browns when using very low silver content. Pure gold and silver were not affected by the irotsuke chemical coating so retained their natural raw colors. Shakudo (shaku 'red' + dō 'copper') was mainly copper with, most often, 3 to 6% gold but could be 10% or more. Gold content can be varied to control the resulting shade of black-blue/black when treated in the traditional chemical bath. The alloys were incorporated using various inlay and overlay techniques and were further worked by carving, engraving, chasing, and polishing [30].

In early antiquity, in very different regions of the world, for example, in the Maikop culture of the Eurasian steppes North of the Caucasus, in Mexico, in the Andes and in Palestine, arsenical copper was used because of its properties which were better than those of Cu and because of its silvery color [31].

1.2   Learn About Silver

To understand better why there were important changes in chemistry that began in the 19th century, we must remember the historical period. The century opens with the companies of Napoleon Bonaparte: his armies entered invasions in many European countries, securing him in a short time the dominion over Europe. His definitive defeat in the Battle of Waterloo (1815) seemed to undo the effects of the Napoleonic era, but Napoleon's work and legacy were far greater and more durable than his military businesses. In the years of glory, he had wiped out the legacy of the Middle Ages: he had abolished feudal rights, restricted the power of the Church, created an efficient administration, developed the economy, and issued a code that constituted the basis of the modern legal organization (Code Napoleon Civil). Napoleon had spread in Europe the conquering of the principles of liberty and equality that had animated the first period of the French Revolution. After the definitive defeat of Napoleon, in 1815 the winning authorities (Austria, Russia, Prussia, and England) met in the Vienna Congress with two goals: restoring the political situation before the French Revolution, creating a situation of territorial balance between the various states; to stifle revolutionary ideals spread throughout Europe. Despite the restoration imposed by the Vienna Congress, between 1820 and 1848, almost all of Europe exploded in various motions, in some cases the revolutionaries asked the monarchs to recognize, through a constitutional charter, greater freedom and rights for the subjects; in other cases, in countries still subjugated to foreign powers, such as Italy, the goal was national independence. Finally, the demands for greater freedom were joined to those of social justice and greater political representation even for the poorest classes. By the middle of the century, Karl Marx (1803–1883), the theorist of socialism accused the bourgeoisie of exercising a true dictatorship by controlling the means of production (industries, machinery, etc.).

The industrial system born in England (industrial revolution) spreads in several countries in different ways in relation to the period and the context in which it occurs. The take-off of Germany, rich in coal and iron, begins relatively later, but after unification, the development is quick and sustained. In the previous decades it was prepared by demographic growth (between 1815 and 1850 the population ranged from 24 to 35 million), from the increase of cultivated land, the abolition of the service, the opening of the borders (Zollverein), the creation of rail and river lines, since the birth of large banks. After the unification, in a few decades, Germany became the largest European steel producer and the leading electricity and chemical company, the three leading sectors of German industry, with large-scale and high-tech companies (Krupp, Thyssen, Siemens, AEG, Bayer, Basf, Hoechst).

These changes also imply a certain order in the sciences, like what happened for chemistry.

After the announcement of the atomic theory in 1803 by English chemist John Dalton, chemists and physicists were called to develop methods for the determination of the atomic masses and give names and symbols to the new elements that were going to be discovered and eventually to arrange all of them according to a given order.

In 1817 the German chemist Johann Döbereneir (1780–1849) became aware of the existence of relationships among elements grouped in triads on the basis of having similar characteristics, for example, lithium, sodium, and potassium; calcium, strontium, and barium; chlorine, bromine, and iodine. However, too many elements were still unknown and Döbereneir could not find any periodicity among the triads on which his law was based.

In the second half of the 19th century the number of discovered elements increased. It became more and more important to look for a unified theory for elements classification and find out if there were a limit to their number. The French geologist Alexandre E. Beguyer de Chancourtois (1820–1886) and the two British chemists John Newlands (1837–1898) and William Odling (1829–1921) separately tried to establish a periodicity by looking at different properties such as chemical for the first and physical for the last two scientists.

No doubt about the contribution of these previous works to the classification of elements into the periodic table in the form as we all know. The modern periodic table had its birth by the discussions at the first International Conference of Chemistry Worldwide in Karlsruhe, Germany, from 3 to 5 September, 1860. At the conference, indeed, one of the main arguments was the method to be used for the determination of atomic masses. The Russian Chemist Mendeleev (1834–1907) and the German Chemist Julius Lothar Meyer (1830–

1895), who both attended the conference, were stimulated to rethink that atomic masses of the elements could be the basis of their classification.

In 1864 Meyer looked for a relationship between the atomic mass and physical properties of the elements and to relate the atomic volume with of the atomic mass, eventually finding a periodic behavior. Meyer disclosed its findings only in 1870, when the Mendeleev Table had already been published, in 1869.

1.2.1 Chemical and Physical Properties of Silver

The tool we currently use to arrange all of the known elements based on their intimate atomic structure, which reflects physical-chemical behavior, is the periodic table of the elements. It is very powerful and conceptually very simple, though. However, many efforts and years of research were needed to find out that the best way to organize elements within the table was the atomic number z, that is, the number of protons found in the nucleus of the atom of a certain element.

According to IUPAC (International Union of Pure and Applied Chemistry), the periodic table is divided into groups (vertical columns, numbered from 1 to 18) and periods (horizontal rows, numbered from 1 to 7). Elements are arranged from left to right and from top to bottom according to their atomic number, z. As that number is linked to electron filling, different blocks can be identified depending on the electron energy sublevel being filled, that is, s, p, d, or f (Fig. 1.3).

The key to elements' periodicity is the way electrons are distributed within energy levels in atoms. In fact, similar distributions can be found in atoms separated by an integer number of periods, that is belonging to the same group.

Due to energy reasons summarized in the Aufbau principle, the orbitals of an atom are filled from the lowest to the highest energy orbitals. Orbitals with the lowest principal quantum number n have the lowest energy, and sublevels energy differences are very small. Nevertheless, this is true only for smaller atoms, that is, for those having $n = 1$ and $n = 2$. For bigger atoms, it happens that some orbitals of a certain level n have higher energy than some orbitals of the subsequent level $n + 1$. As a consequence, the electron filling process proceeds zigzagging through the atom orbitals as can be seen in Fig. 1.4.

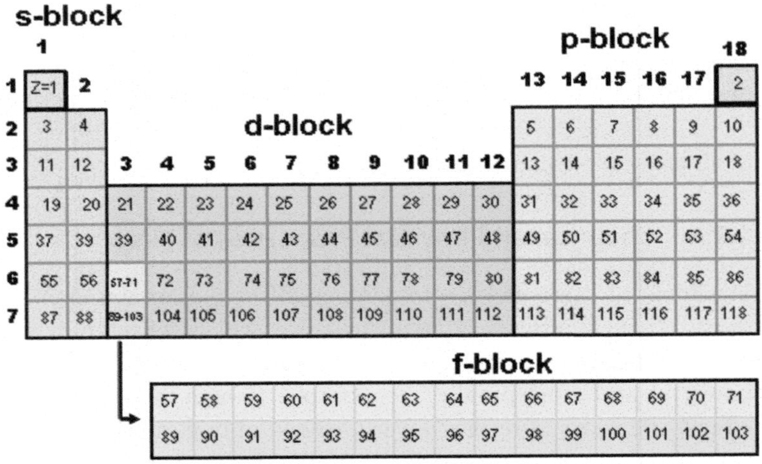

s-block

p-block

1	2	3	4	5	6	7	8	9	10	11	12	13	14	15	16	17	18
1 Z=1	2																2
2 3	4				d-block							5	6	7	8	9	10
3 11	12	3	4	5	6	7	8	9	10	11	12	13	14	15	16	17	18
4 19	20	21	22	23	24	25	26	27	28	29	30	31	32	33	34	35	36
5 37	39	39	40	41	42	43	44	45	46	47	48	49	50	51	52	53	54
6 55	56	57-71	72	73	74	75	76	77	78	79	80	81	82	83	84	85	86
7 87	88	89-103	104	105	106	107	108	109	110	111	112	113	114	115	116	117	118

f-block

57	58	59	60	61	62	63	64	65	66	67	68	69	70	71
89	90	91	92	93	94	95	96	97	98	99	100	101	102	103

Figure 1.3 Scheme of the periodic table of the elements with periods and groups designation from 1 to 7 and from 1 to 18, respectively. Atomic number, z, in each box increases from left to right and from top to bottom.

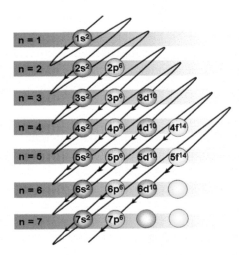

Figure 1.4 Atomic orbitals with an indication of the possible maximum number of electrons in each sublevel s, p, d, and f. The zigzagging arrow indicates the increase of atomic number z.

Starting from the second one, each period begins with an element of group 1, characterized by atoms with external electron configuration ns1 and ends with an element of group 18, characterized by atoms with external electron configuration ns2 np6, where n is the principal quantum number corresponding to the number of the period as well. These two groups are commonly called alkali metals and noble gases, respectively. Some other trivial names of old origins are still in use such as alkaline earth metals (group 2), chalcogens (group 16), halogens (group 17), and rare earth elements (group 3 including most of the f-block elements).

Each group comprises elements which atoms have the same external electronic configuration while dimensions increase as the period number increase. The chemical properties of a group exhibit similarities because the outer electrons are precisely those that influence the chemical processes.

The elements corresponding to electron filling of d-orbital levels are commonly called with the name of transition elements. For the above-cited principle about energy levels of atoms, it is the orbital nd that is being filled in the $n + 1$ period for the transition elements. In fact, the energy of 3d orbital is higher than 4s, that of 4d is higher than 5s and so on.

With our up-to-date scientific knowledge, it is quite easy to understand why copper, silver, and gold are part of the same group of the periodic table. It is therefore even more remarkable to observe that some ancient alchemists, among all the elements they knew, did relate copper to silver and gold indeed.

Within the periodic table, silver shares with gold and copper the group 11. Copper, silver, and gold are part of the transition elements series, being their orbitals 3d, 4d, and 5d, respectively, in the course of filling.

Due to the mentioned energy reasons, the elements of group 11 have a single ns1 electron external to a completed $(n - 1)d^{10}$ shell, instead of $(n - 1)d^9 ns^2$ as it would be expected, as reported in Table 1.4.

These elements might therefore easily assume the +1 oxidation state in a similar fashion to alkali metals that have similar ns^1 external electron configuration. Nevertheless, they can assume oxidation states higher than +1 when d electrons are involved beside s ones as summarized in Table 1.5.

Table 1.4   Electron configuration of Group 11 elements at oxidation state 0

Electron Shell (n)	1	2	3	4	5	6
Quantum Subshell (l)	s	s p	s p d	s p d f	s p d f	s
Cu	2	2 6	2 6 10	1		
Ag	2	2 6	2 6 10	2 6 10	1	
Au	2	2 6	2 6 10	2 6 10 14	2 6 10	1

Table 1.5   Electron configuration of Group 11 ions at different oxidation states

External configuration			
Cu	Ag	Au	Oxidation state
$3d^{10}\,4s^1$	$4d^{10}\,5s^1$	$5d^{10}\,6s^1$	0
$3d^{10}$	$4d^{10}$	$5d^{10}$	+1
$3d^9$	$4d^9$	$5d^9$	+2
$3d^8$	$4d^8$	$5d^8$	+3

As mentioned earlier, the elements of the same series have many physical and chemical properties in common because they have only slight differences in electron configuration and therefore, in turn, the values of atomic radii, ionization potential, and electronegativity to cite a few. However, each element possesses its own unique features. In Table 1.6, are summarized some chemical and physical properties of silver, practically its identity card.

All transition elements are metals, and many of them have catalytic properties because of the possibility to switch from one energy state to another with small energy variations. The transition metal ions contain an incomplete d shell of electrons, and the color is due to electronic transitions involving the d electrons. The presence of incomplete d orbitals, and the high electric field of the ions are the reason because of these elements can easily give rise to coordination compounds or complexes (see Section 1.2.4).

As it can be expected, copper, silver, and gold have some physical and chemical similarities. All of them have high melting point (about 1000°C), they are ductile and malleable, they have rather high density and high electrical conductivity. In nature they occur also in native state in nature and are obtained by reducing their compounds without difficulty. Silver and gold are not oxidized by air, only copper is slowly oxidized. Furthermore, their chloride salts are highly insoluble in water. The crystal structure is face-centered cubic system for all.

Table 1.6   Summary of chemical and physical properties of silver filling

Atomic Number	47
Atomic Weight	107.868
Ionic Radius (Å)	1.442
Oxidation State	1, 2, 3
Standard Electrode Potential at 25°C (V)	Ag_2O/Ag 0.342 AgO/Ag_2O 0.599 Ag_2O_3/AgO 0.740 Ag^+/Ag 0.799 Ag^{++}/Ag^+ 1.987
Ionization Energy	7.87 (eV)
Melting Point	961.93°C
Boiling Point	2212°C
Heat Capacity	0.0558 (cal/g/°C)
Thermal Conductivity at 20°C	1.02 (cal/sec/cm/cm^2/°C)
Electrical Resitivity at 20°C	1.59 ($\mu\Omega$/cm)
Hardness (Mohs Scale)	2.5–3

Copper, silver, and gold all form metallic bonds among atoms. Dissimilarly to covalent and ionic bonding, metallic bonding is non-directional; such a strong bond is due to positively charged metal atoms, in fixed positions, surrounded by what can be figured as a sea of delocalized electrons. This explains the high electrical conductivity and high ductility. Furthermore, the tight packing of atoms promotes the transmission of heat giving good thermal conductivity.

Silver, gold, and copper are diamagnetic with absolute permeability slightly greater than that of vacuum. In the absence of magnetic field "magnets" are random, and there is no magnetic effect. In the presence of an external magnetic field there is a weak attractive force and the dipoles tend to align along the lines, even if at room temperature the thermal agitation counteracts the diamagnetic orientation.

In nature, we can find two stable isotopes of silver: ^{107}Ag and ^{109}Ag, the first of which is the most abundant (51.83%) [32].

The isotope ^{107}Ag, was produced in the early "years" of the solar system by the rapid decay of ^{107}Pd. The latter is so unstable that it decayed within the first 30 million years of solar system history. Silver and palladium differ in their physico-chemical properties being silver is more volatile while palladium binds to iron more likely. These differences allowed the researchers of Carnegie University to use the isotope ratios in primitive meteorites and mantle rocks to relate the history of the volatile elements of the Earth and to the formation of the iron core of the planet [33]. Other evidence relating to indium and tungsten isotopes indicates that the core formed between 30 and 100 million years after the origin of the solar system.

Silver isotopes suggest that the Earth's core formed about 5–10 million years after the birth of the solar system, and that is earlier than indicated by the data for the hafnium-tungsten. Researchers thus concluded that the accretion process of the Earth to have begun with materials free of volatile elements until it reaches 85% of its final mass; they were then scanned in later stages of formation, about 26 million years after the origin of the solar system, probably during a single event. Perhaps the giant collision between the proto-Earth and an object the size of Mars has also projected in its orbit enough material to form the moon.

The results of the study corroborate a 30-year-old terrestrial formation model called "heterogeneous accretion," according to which the primary components of the Earth changed in composition gradually that the planet was forming. According to the researchers, it would have sufficed a small amount of volatile-rich material like primitive meteorites elements added during later stages of Earth's accretion to account for all the volatiles, including water, currently on our planet.

Geochemistry has as its object the determination of absolute and relative abundance of chemical elements in the Earth, their distribution and migration over time. The term was introduced for the first time, by the German chemist C. F. Schönbein (1799–1868) in 1838. Chemistry and physics studies leading to the discovery of optical emission spectroscopy, crystal X-ray diffusion, and natural radioactivity have greatly influenced the development of geochemistry. The fundamental objectives of this science are the determination of the abundances of the elements (and of the nuclides) in the various units of which the Earth is constituted, the formulation of the laws that regulate the migration and the fractionation of the elements and of the isotopes in natural environments, the study of chemical-energetic changes connected with geological processes. The fundamental objectives of this science are the determination of the abundances of the elements (and of the nuclides) in the various units of which the Earth is constituted, the formulation of the laws that regulate the migration and the fractionation of the elements and of the isotopes in natural environments, the study of chemical-energetic changes connected with geological processes. Through the measurement of quantities that are dependent on time, geochemistry finally reaches the quantitative evaluation of geological time. The calcophilic elements are those metals and not metals heavier that possess a low affinity for oxygen and prefer to bind with sulfur to form highly insoluble sulfides, silver is called calcophile. Due to the fact that these sulfides are much denser than the silicate minerals formed by the lithophile elements, the calcophiles separated underneath the lithophiles layer during the period when the Earth's crust was formed. This caused their depletion in the Earth's own crust relative to their abundance in the solar system, although thanks to the formation of non-metallic minerals this depletion did not reach the levels found for the siderophile elements. transition metals that possess high density and tend to bind with the metal iron both in the solid and in the molten state [34].

1.2.2 Silver Extraction and Recovery Methods

Within elements of group 11 of the periodic table, silver is a rare element of the Earth's crust accounting for about 0.075 parts per million (ppm) overall. Gold is even more rarer (approx. 0.002 ppm) while copper is definitely more (approx. 60 ppm). Natural silver is very widely distributed as a main component of some rare minerals or as a contaminant of some largely diffused ones.

Sources of silver can be grouped into two main clusters depending on the origin of the material to be extracted that is natural (primary sources) or from recycling (secondary sources). Primary sources are those where silver can be extracted from dedicated mining fields or as a by-product of different non-ferrous metals treatments and refining including those of noble metals like gold. In secondary sources silver can be recovered from metal scraps, jewelry, fine chemicals industry wastes, old photosensitive films, microelectronics, etc.

Most of the silver comes from primary sources [35]. The mineral containing the highest percentage of silver is *argentite*, a silver sulfide enriched ore that can contain up to 87% of silver. The name *argentite* refers to cubic silver sulfide stable over 177°C, below this temperature it converts to *acanthite*, orthorhombic.

Modern extraction methods of silver are still based on centuries-old techniques such as roasting, alloying, and chemical leaching. Processes of recovering silver as a by-product during the production of non-ferrous metals, typically lead, zinc, and copper, or noble metals, is usually followed by electrochemical refining (Fig. 1.5).

Figure 1.5 Schematic diagram of silver production.

In the beginning, the only extracting technique of silver was to smelt silver-enriched crushed rocks in oxidizing or reducing conditions to obtain silver, while base metals were discarded as waste. Starting materials were almost exclusively those ones containing silver in elemental state or as salts, sulfates, sulfides, and chlorides mainly.

Just less ancient methods based on mercury extraction have been in use for several centuries. After ore treatment with mercury to form an alloy with silver (amalgamation), the precious metal is recovered from amalgam by mercury evaporation. Nowadays amalgam processes have definitively lost attraction when in large-scale operations are involved due to high costs, low efficiency, and mercury toxicity. On the other hand, they are still in use for small scale productions.

Amalgam processes were progressively abandoned to adopt, around 1800, those based on thiosulphate or metallurgy. The first one was effective and cheaper than amalgam methods, however it was in turn replaced by cyanide based ones, which are still in use also for gold. Metallurgical methods are favorably applied when the metals accompanying silver in the starting material are functional to the success of the process, as in the case of copper and lead.

Although mining is a very old business, the world's largest silver mine is a quite recent discovery. In Queensland, Australia, in 1990 it was opened a mine which capacity has reached 3 million tons of ore per year. Other important active big mines can be found in Mexico, Poland, Bolivia, and Turkey. Silver mining happens to be a worldwide activity nowadays, probably more spread than in the past, when the technology was rougher. Going back to ancient times, at least as long as we know it is the beginning of documented exploitation of silver mines, Ancient Egypt silver came from mines in Asia Minor. At the time of Romans, when silver became so widely used as currency, it probably came mainly from Britannia [36].

An interesting part of the research project "The Demes of Attica and the Peloponnese War. Community Resilience" directed by Ghent University (Belgium), led to the discovery of an incredible mine in Thorikos (Greece). Characterized by about 5 km of underground galleries, it represents one of the largest mines found on the Aegean Sea and dates back to the Late Neolithic/Early Helladic, around 3200 BC. Archaeologists have found lamps, and pots that testify the activity of the period.

Others ancient mines were located in the Iberian Peninsula, in Anatolia, and in the Kosmaj region (mountain south of Belgrade, the capital of Serbia).

In the ancient world, the smelting technique called cupellation was one of the most diffused methods for metal extraction. The origin of the name comes from the Latin *cupella*, meaning jar, barrel. The cupellation is an effective metallurgy process that was used to refine noble metals such as gold and silver by separating from other metals present in the natural occurring ore or alloy, the main components of which are lead, copper, zinc, arsenic, bismuth, and antimony. The process is based on the principle that precious metals do not oxidize or chemically react, unlike what happens to other metals; so, when heated to high temperatures, the precious metals are separated while the others react to form slag or other compounds. Slag dumps found in archaeological campaign can be the sign of ancient metallurgy activities. Cupellation on a small scale was also used for assaying noble metals content in ore or alloys (Fig. 1.6).

ROUND ASSAY FURNACE.

RECTANGULAR ASSAY FURNACE.

Figure 1.6 Cupellation furnaces [37].

Silver was mostly extracted from argentiferous galena (lead sulfide) or cerussite (lead carbonate) were it is present in small amounts—a few percent units—as corresponding salt or solid dispersion. The ore was placed in a crucible then introduced into a furnace where the temperature was appropriately raised up to about 1000°C while maintaining an oxidizing environment by means of air insufflation. Sulfur and carbon that are part of the anions of the mineral salts oxidized to volatile compounds (sulfur and carbon dioxides) and removed. At the same time, lead formed melted litharge (lead oxide, melting point 888°C) that was drained by absorption on the cupel. The noble metal—that does not react with oxygen—was eventually recovered after cooling at the bottom of the crucible as a bead [38].

Despite the straightforwardness of the cupellation process, many reactions are involved. For example, let consider a sulfide of generic formula MeS that could be a matrix of the argentiferous ore. Inside the furnace, a number of chemical equilibria can be established as depicted in Fig. 1.7. In practice, by supplying an excess of oxygen in the second stage, and getting rid of sulfur dioxide, it is possible drive the process toward the formation of MeO. On the other hand, under these conditions silver that can be present as is or by first stage reduction and gold do not form oxides or salts, thus allowing their recovery as a pure material (Fig. 1.8).

Figure 1.7 Schematic diagram of silver production by cupellation.

$$2Me + O_2 \leftrightarrows 2MeO$$

$$2MeS + 3O_2 \leftrightarrows 2MeO + 2SO_2$$

$$2MeO + 2SO_2 + O_2 \leftrightarrows 2MeSO_4$$

$$MeS + 2O_2 \leftrightarrows MeSO_4$$

$$Me + SO_2 \leftrightarrows MeS + O_2$$

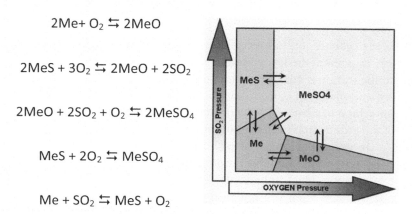

Figure 1.8 (Left) Example of reactions involving base metal during the treatment of argentiferous material when submitted to cupellation. (Right) Schematic phase diagram of species at equilibrium under constant high-temperature conditions and variable partial pressures of sulfur dioxide and oxygen.

In addition to the goal of producing silver by extraction of ore, cupellation principles were successfully applied for the qualitative and quantitative determination of silver content in ore, the so-called fire assay [39]. It is worth noting that this ancient practice was developed and successfully applied long before knowing the principles on which the pyro-metallurgical technique relies on, just because human beings have always been good observers of nature.

Definitely more recent methods to extract precious metals from their low content ore are based on the selective leaching by means of suitable chemicals. For example, silver, that is generally present in minerals as chloride, sulfide, and carbonate, can be successfully extracted, even when in very low amounts, by cyanide ion. The cyanide process—also known as Macarthur-Forrest process after the names of its developers (1887)—is based on the reaction among silver salts and cyanide ions to form a very stable ion complex in a water solution from which metallic silver is successively recovered by exchange reaction with metallic zinc.

In practice, on a small scale, the ore containing the silver salt is finely grounded then suspended in an adequate excess of potassium cyanide (KCN) aqueous solution through which air is bubbled. The reactions of silver chlorides and sulfides are the following:

$$AgCl_{(solid)} + 2\ KCN_{(aq)} \rightarrow [Ag(CN)_2]^- K^+_{(aq)} + KCl_{(aq)} \qquad (1.3)$$

$$Ag_2S_{(solid)} + 4KCN_{(aq)} \rightarrow 2\ [Ag(CN)_2]^- K^+_{(aq)} + K_2S_{(aq)} \qquad (1.4)$$

Silver cyanide ion is a very stable linear complex in which silver ion is present in its oxidation state +1. The external electronic configuration is $4d^{10}$, being the $5s^1$ electron absent, that is, d orbitals are completely full. Coordination bonds involve one 5s- and one 5p-orbital of silver (sp-hybridization) yielding a linear arrangement of the groups that form the complex ion.

Oxygen presence can help shifting the equilibrium toward silver cyanide ion complex formation because of sulfur anion removal by oxidation. Most importantly, oxygen is necessary to oxidize the noble metal when present in its metallic state or metal alloys to make it available for leaching, as it is more common for gold rather than silver:

$$4Au + 8KCN + O_2 + 2H_2O \rightarrow 4KAu(CN)_2 + 4KOH \qquad (1.5)$$

The same type of process can be applied to obtain silver and gold from low content precious metal minerals by treating remarkable amounts of materials as those obtaining after mining and concentration [40].

Eventually, silver recovery is obtained by an exchange reaction with zinc. In practice, the latter is added to a solution of silver cyanide ion to cause silver reduction and subsequent precipitation:

$$2\,[Ag(CN)_2]^- + Zn \rightarrow [Zn(CN)_4]^{2-} + 2\,Ag \downarrow \qquad (1.6)$$

After being extracted, silver needed to be worked up to obtain a quality grade suitable to meet the requirements of different industries demand. The cyanide method generates precious metals of good quality; however, they are usually contaminated by other metals thus cannot be used as they are for specialty uses. High-purity silver can be obtained by applying electrochemical methods to silver ions solutions, as for example electrolysis of aqueous silver nitrate ($AgNO_3$) often added with copper nitrate ($CuNO_3$) to increase the conductivity of the bath.

As it happened in various fields, a particular waste or even a pollutant can become a source of worth materials. Secondary sources of silver, and other rare metals, are becoming more and more important due to the wide diffusion of silver containing high value-added goods. In fact, after being extracted, silver is worked up to obtain a quality grade suitable to meet the requirements of different industries demand. Used manufactured goods containing silver and other rare metals, at the end of their lifecycle, can be sent to the recycling industry where valuable metals are recovered. From this practice, some of the costs of refining are gained back, as it can be expected. However, and most importantly, significant benefits also involve environmental protection.

In the first steps of the recovery process the spent electronics is crushed, washed and dried then suitably treated to dissolve the material of interest. The chemical treatment to be applied is a major challenge because the process should be efficient and cost effective while taking into account the impact of the matrix that must be removed (plastics and other metals). Mixtures of sulfuric acid and thiourea are of common use. These allow a silver extraction up to 98% of the actual content yielding silver ion solutions. A largely applied technique to recover silver from extraction solutions is the cementation, that is, an electrochemical process capable of yielding high-purity metals by reductive process. In practice, a sulfuric or nitric silver solution is put in contact with a metal, commonly zinc or copper or manganese, to obtain metallic silver by chemical exchange. It is possible to chemically reduce silver ion to metal also by cellulose, hydrazine, and ascorbic acid.

The removal of pollutants, including heavy metals, by biomass absorption (biosorption) is cheap process easily applicable in various fields, also on a large scale, that is the subject of continuing research. Biosorption based processes are particularly attractive when considering both environmental and economic aspects. In a similar fashion, silver ions can be recovered from aqueous solutions, that are not necessarily waste, by the action of biomasses consisting of particular bacteria, algae, fungi, etc. selected to maximize the extraction rate.

Ion exchange resins can be suitably used in a similar way as biosorption for environmental purposes. Also, in this case the principle can be applied to silver ions solutions for recovery and recycle purposes. When aqueous solutions are contacted with the resins, packed in columns or in tanks, silver ions are fixed to the resin while the original ions are released from the resin. Silver is eventually eluted by washing with solutions of N, N-dimethylformamide, thiosulphate, strong acids, complexants, etc. from which solutions it can be recovered by the usual methods (Fig. 1.9) [41].

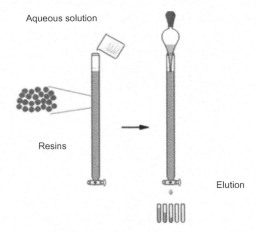

Figure 1.9 Silver recovery by ion exchange resins.

An interesting method for silver and gold recovery from printed circuit boards of waste electric and electronic equipment (WEEE), more specifically addressed to mobile phones, is based on the use of acidic thiourea coupled with a sorbent gel derived from persimmon tannin [42]. After mechanical crushing, the WEEE is heated at high temperature to eliminate plastics. The residue is finely grinded then treated with a solution of thiourea in sulfuric acid. After that, the extract solution is mixed to the sorbent gel obtaining a mixture that can be easily incinerated in the end. The main advantage is the low toxicity of employed chemicals compared to the cyanide method.

Minute batteries that are widely used in a number of small electronic devices are built around the electrochemical couple Ag_2O/Zn. Such button cells—as they are commonly called because of their usual shape—typically weight from 0.2 g to 2.3 g and contain about 20% by weight of silver equivalent.

In spent cells, silver is mainly found as a metal along with a variable residual amount of the original oxide depending on the cell use. In this particular case, the environmental aspects are as important as economics or even more important due to presence of mercury traces. Process recovery of silver from button cells encompasses crushing followed by oven drying and sieving to remove non-metallic parts that, in this case, are very few. After material homogenization, the precious metal is extracted by means of diluted aqueous nitric acid which dissolves silver oxide and oxidizes metallic silver to silver ion:

$$3Ag + 4HNO_3 \rightarrow 3AgNO_3 + 2H_2O + NO \qquad (1.7)$$

Subsequently, silver ion is precipitated as a chloride salt, which is very insoluble in acidic conditions, by adding potassium chloride to the nitric solution. More than 99% of the original silver content can be recovered as silver chloride [43]. It is worth noting that mercury remains in the nitric solution and it can be successively recovered by using a suitable ion exchange resin. In the end metallic silver can be obtained by exchange reaction with zinc in the presence of hydrochloric acid.

Silver is also present in sea water at very low concentrations that is about 10 parts for billion (0.01 ppm) [44]. So far, it cannot be considered a possible source of the precious metal, however that concentration value should be regarded with due attention because its fluctuation in sea water is a direct index of pollution. In fact, if silver is present in coastal zones at relatively higher concentrations marine organisms and the whole costal ecosystem can be damaged.

Silver and chemical compounds containing silver are employed by various industries for different processes and applications. Silver is mostly recycled, or at least recycling should be the goal, however small portions of this metal are non-intentionally released into the environment as a waste by-product.

Industrial sources of silver as a pollutant were much larger in the past than they are today, and their relative impact changed over the times according to silver-based goods demand. Anyway, notable silver sources include emissions from the photographic and electrochemical plating industries, mining extractive activities, urban refuse, sewage treatment plants, specialty metal alloy production, and electrical components.

Standards were established for industry and laboratory disposal practices of drain water into sewer systems, safe silver limits in the public water supply, and thresholds of adverse effects of silver on the biosphere.

1.2.3 Silver Compounds

The most common state of oxidation of silver is +1, due to the loss of one electron. Likewise, copper and gold, it can also assume the less common oxidation states of +2 and +3. Silver can easily form stable compounds by bonding with other different elements such as sulfur, oxygen, and halogens.

With oxygen, silver constitutes the most thermodynamically stable oxides Ag_2O and Ag_4O_4 [45].

The first one is used as a material to build the cathode for the aforementioned button cell batteries, based on the couple Ag_2O/Zn. During the discharge, the zinc of the anode oxidizes to ZnO and the Ag_2O of the cathode reduces to Ag while supplying a nearly constant operating potential difference of about 1.5 V:

$$Zn + Ag_2O \rightarrow ZnO + 2Ag \qquad (1.8)$$

Because the good power properties and ease of miniaturization Ag_2O/Zn button cell batteries are widely employed as the power source for many small electronic devices like wrist watches, pocket calculators, artificial cardiac pacemakers, implantable cardiac defibrillators, hearing aids, etc.

The second oxide is less ordinary, and the oxidation state is only apparently +2. Indeed, Ag_4O_4 is a mixed oxide where silver is present with oxidation state +1 and +3. In other words, Ag_4O_4 that is an equimolar mixture of Ag_2O and Ag_2O_3. Due to its higher oxidant power than Ag_2O, it has a potential for its usage in medical fields [46]. It can be used to treat several skin diseases such as eczema, dermatitis, and diabetic foot ulcers.

The most common usage of Ag_4O_4 is as a cathode material for zinc-silver oxide button cell batteries, which provide the power source for many small electronic devices. These primary cells are characterized by a high energy output per unit weight, high discharge at approximately constant voltage (~1.5 V) and excellent shelf stability at room temperature. The performance of $Zn–Ag_4O_4$ batteries deteriorates considerably with operation at elevated temperatures (above 323 K), due to internal complications arising from the thermal decomposition of the Ag_4O_4 cathode component [47].

During World War II, the mixed oxide Ag_4O_4 was used in the filters of gas masks for protection against carbon monoxide. The toxic gas is easily oxidized and sequestrated because of the concurrent reduction of silver from +3 of one component of the mixed oxide to +1 of silver carbonate:

$$Ag_4O_4(s) + 2CO \rightarrow 2Ag_2CO_3 \qquad (1.9)$$

The mixed oxide Ag_4O_4 can be obtained by chemical synthesis, via persulfate oxidation of a soluble silver salt, such as sulfate, in alkaline media as depicted in the following reaction [48]:

$$2Ag^+ + S_2O_8^{2-} \rightarrow 2Ag^{2+} + 2SO_4^{2-} \qquad (1.10)$$

$$4Ag^{2+} + 8OH^- \rightarrow Ag_4O_4 + 4H_2O \qquad (1.11)$$

Ozone interacts with silver powder giving rise to different oxides like Ag_2O and Ag_4O_4. The mixed oxide Ag_4O_4 is formed at the gas/oxide interface via the oxidation of Ag_2O [49] according to the following reactions:

$$2Ag_{(s)} + O_{3(g)} \rightarrow Ag_2O_{(s)} + O_{2(g)} \qquad (1.12)$$

$$2Ag_2O_{(s)} + 2O_{3(g)} \rightarrow Ag_4O_{4(s)} + 2O_{2(g)} \qquad (1.13)$$

An interesting new methodology for Ag_4O_4 preparation by means of a physical technique such as pulsed laser deposition (PLD) has been developed at the Micro and Nanostructured Materials Laboratory at the Politecnico di Milano where I work, and it has been object of a patent application [50] by means of a physical technique such as pulsed laser deposition, whose acronym is PLD. The obtained product is of high purity, nanostructured with size control and high antibacterial activity. The weakness of the method is the difficult scalability at the industrial level.

Silver sulfide Ag_2S, is a stable black compound that has three polymorphic crystal structures depending on temperature. At low-temperature we can find monoclinic phase α-Ag_2S (acanthite) existing at temperatures below ~177°C; β-Ag_2S phase (argentite) with body centered cubic (bcc) sublattice of sulfur atoms exists in the temperature range 179–586°C; and at high-temperature face-centered cubic (fcc) phase γ-Ag_2S stable from 587°C to 825°C (melting temperature) [51]. For practical application as a semiconducting photoluminescent material, of most interest is low-temperature monoclinic α-Ag_2S (acanthite) phase in nano-sized state. However, the crystal structure of this phase was determined mostly on natural samples of acanthite mineral. Silver jewelry darkens not for oxidation by oxygen but for sulfide action.

Among silver halides, bromides and chlorides are of particular importance due to their photography related use. Despite the diffusion of digital cameras, as such or embedded in smartphones and the like, relegated the silver-based photography to a niche, silver halides production is still an important market. Silver chloride occurs in nature as cerargyrite or horn silver minerals.

Silver is employed in dentistry for its malleability and its non-toxicity in alloy with other metals, but the combination with mercury is not used yet [52].

Silver is not always a quiet salt; its salt with fulminic acid (HCNO), an isomer of isocyanic (H-NCO) acid, is extremely dangerous because of the high tendency to self-explode. The suffix fulminate derives from the Latin word *fulmen* which means flash. Silver fulminate (Ag-CNO) was discovered not intentionally by the chemist Luigi Valentino Brugnatelli in 1800. Due to its very high explosive tendency and sensitivity to friction this compound was used in detonators [53].

Silver nitrate is a water-soluble salt obtained by treatment of silver as such or its compounds with diluted or concentrated nitric acid. As a solid it should be handled with care as it is very corrosive, moreover it generates dark spots on surfaces, including skin, if left on exposed to light. Therefore, the use of gloves and to clean work surfaces is mandatory to avoid problems. Silver nitrate is the main ingredient of a topical medication used to cauterize skin and to remove warts that is applied by a dedicated device called caustic pencil or silver nitrate stick.

1.2.4 Silver Coordination Compounds

The high electric field of transition metal ions and the presence of incomplete d orbitals gives them the ability to give rise to coordination compounds.

Coordination compounds (complexes were an old definition, still used by chemists) are entities that have a metal center that is bound to atoms or molecules electron donors, named ligands. Coordination compounds can be neutral or charged; when they are charged, must be stabilized by neighboring counter-ions.

In nature coordination compounds are indispensable to living organisms. In biological systems we can find a variety of metal complexes that play important roles.

We have metalloenzymes that regulate biological processes; carboxypeptidase is a hydrolytic enzyme important in digestion, contains a zinc ion coordinated to several amino acid residues of the protein. Catalase is an enzyme that is efficient catalyst for the decomposition of hydrogen peroxide, it contains iron–porphyrin complexes. In both cases, the coordinated metal ions are probably the sites of catalytic activity.

Hemoglobin is a protein specialized in oxygen transport, it is found inside the red blood cells to which it gives the characteristic red color. It is a globular protein composed of four protein chains; each chain contains a prosthetic group called *heme* formed by a porphyrin ring that binds to the center an iron atom in the form of a positive ion Fe^{2+} (Fig. 1.10).

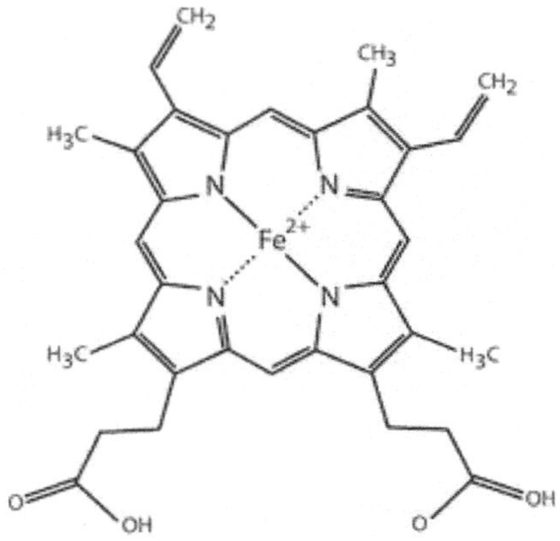

Figure. 1.10 Heme group.

The iron atom in the center of the porphyrin ring binds the oxygen molecule reversibly and transports it to all the cells of our body through the blood.

Hemoglobin can also bind and transport other molecules such as nitric oxide (NO), carbon monoxide (CO), and cyanide (CN⁻). Nitric oxide is active on the wall of blood vessels by regulating the arterial pressure. Instead carbon monoxide and cyanide are very toxic molecules because they bind to the iron of the heme in an irreversible way, preventing the normal bond with oxygen and therefore its transport to the tissues. Death occurs by asphyxia.

Chlorophyll, a magnesium-porphyrin complex, and vitamin B12, a complex of cobalt with a macrocyclic ligand known as corrin, are other important coordination compound for living organisms (Fig. 1.11).

Chlorophyll is the pigment present in all plants, responsible for the green color, has the purpose of absorbing sunlight and triggering photosynthesis, the process by which carbohydrate are produced starting from water and carbon dioxide.

$$6CO_2 + 6H_2O + h\nu \rightarrow C_6H_{12}O_6 + 6O_2 \qquad (1.14)$$

Chlorophyll is essential for life to such an extent that plants that do not produce chlorophyll, such as mushrooms, survive only by planting themselves on the roots of plants with green leaves or using the humus produced by the decomposition of green leaves. Chlorophyll is also very important for our body: its chemical composition recalls that of hemoglobin. In the case of chlorophyll, the central iron atom present in the hemoglobin is replaced by magnesium.

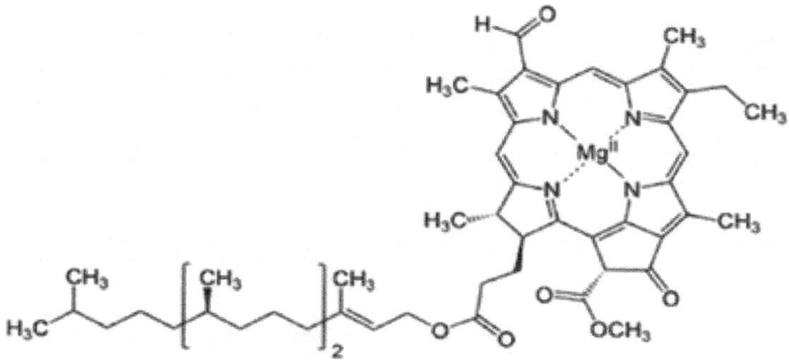

(a)

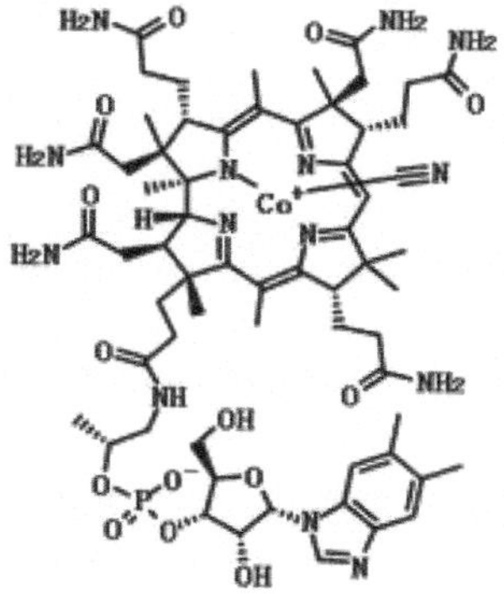

(b)

Figure 1.11 Structural formula of (a) chlorophyll and (b) vitamin B12.

Vitamin B12, or cobalamin, is a fundamental vitamin in the processes of red blood cell formation and for the cells of the nervous system. Together with folic acid—another B group vitamin—it supports the synthesis of DNA and RNA. In nature, it is produced by some bacteria that humans have learned to cultivate *in vitro* and which they are able to use to produce special supplements. It is the most complex of the eight types of vitamin Bs and is water-soluble, that is, a vitamin dissolving in water. Vitamin B12 is also essential to prevent a very dangerous form of megaloblastic anemia, known as pernicious anemia.

In chemistry and technology there are many examples of coordination compounds. One of the most known coordination compounds is iron (II, III) hexacyanoferrate(II, III) or ferric ferrocyanide. Prussian blue $KFe[Fe(CN)_6]$, a color used by randomly obtained artists for heating between animal waste and calcium carbonate Na_2CO_3 in an iron vessel. There are known for a long time also potassium hexacyanoferrate(II) $K_4[Fe(CN)_6]$ (1753), potassium hexachloroplatinate(II) K_2PtCl_6 (1760–1765) and hexaamminecobalt(III) chloride $[Co(NH_3)_6]Cl_3$ (1798).

In 1798 the French chemist B. M. Tassaert (1765–1835) obtained orange crystals of $[Co(NH_3)_6]Cl_3$, leaving a mixture of cobalt chloride $CoCl_2$ and aqueous ammonia NH_4OH. He was baffled by the fact that two stable substances could be combined, and form a new product, with characteristics completely different from those of the two components.

The Swiss chemist Alfred Werner (1866–1919) introduced at the beginning of the 20th century the idea of primary and secondary valence to explain this phenomenon. Valence may be defined like the maximum number of univalent atoms (originally hydrogen or chlorine atoms) that may combine with an atom of the element under consideration, or with a fragment, or for which an atom of this element can be substituted. Werner identified *Hauptvalenz* primary or ionizable valence, and *Nebenvalenz* secondary or non-ionizable valence. The first represents the oxidation state, the second represents coordination number. The oxidation state provides the degree of oxidation of an atom in terms of counting electrons; while coordination number of a specified atom in a chemical species is the number of other atoms directly linked to that specified atom He also demonstrated that the determining factor for the chemical characteristics of the coordination compounds was not the primary valence, but the number of secondary valences possessed by the metal ion. The Nobel Prize in Chemistry 1913 was awarded to Alfred Werner "in recognition of his work on the linkage of atoms in molecules by which he has thrown new light on earlier investigations and opened up new fields of research especially in inorganic chemistry" [54].

In coordination compounds, the central metal is able to accept one or more electronic doublets from a donor atom that very often forms part of a molecule.

IUPAC has established some rules for the nomenclature of coordination compounds [55].

For brevity we report only some rules: (a) the cation is named before the anion; (b) non-ionic or molecular complexes are called with names formed by a single word; (c) the binders in a complex are indicated in order (1) negative, (2) neutral, (3) positive, without separating lines; (d) prefixes bis-, tris-, tetrakis-, etc. are used in front of complex names (especially containing the prefixes mono-, di-, tri-, etc. in the name of the same ligand); (e) the oxidation state of the central atom is marked by a Roman numeral in brackets at the end of the complex's name, with no spacing; (f) for a negative oxidation state, the minus sign is placed in front of the Roman numeral, 0 is used for the zero oxidation state; (g) geometric isomers are generally indicated by the terms cis to indicate adjacent positions (excluding positions at 90°) and trans for indicating opposite positions (excluding 180°).

The most common coordination numbers are 2, 4, and 6; the geometry can therefore be planar, tetrahedral, or octahedral.

The donor molecule (or ion) is called a binder. Some examples of binders are NH_3, H_2O, Cl^-, *en* (ethylenediamine), EDTA (ethylenediaminetetraacetic acid), and with aromatic molecules such as bipyridyl and *o*-phenanthroline. The resulting complex can be positive, negative, or neutral.

The binder can be monodentate in case it has only one donor atom or multidentate in case it has more than one donor atom. It is said that the coordinated multidentate ligands are chelated, the term derives from the Greek χηλή, which means pincer.

Silver (I) has coordination number 2 and forms coordination compounds preferably with nitrogen, phosphorus, and sulfur (Table 1.7).

There are also some complexes with silver (II); the reason is found in the high value of the potential required to oxidize silver (I) to silver (II):

$$Ag^{2+} + e^- \rightarrow Ag^+, E°=1.98 \text{ V} \tag{1.15}$$

There are complexes with nitrogen bases such as pyridine, dipyridil, and *o*-phenantroline [56].

Regarding silver (III) there two complexes obtained by oxidation with persulfate of a mixture of silver sulfate Ag_2SO_4, soda NaOH and tellurium dioxide TeO_2 in water. The resultant complexes are $Na_6H_3[Ag(TeO_6)_2]\cdot18H_2O$ and $Na_7H_2[Ag(TeO_6)_2]\cdot14H_2O$.

Table 1.7 Silver coordination compounds

Silver complexes	Notes
With ammonia	Complexes with ammonia are very stable. Common silver salts that are insoluble in water all dissolve more or less readily in aqueous ammonia.
With halogens	Complex ions with bromide and iodine are more stable than those with chlorine.
With cyanides	Dicyanoargentate $[Ag(CN)_2]^-$, is the complex most generally recognized; we have seen above the important process of cyanidation.
With thiocyanate	Thiocyanate $(SCN)^-$ is another potential lixiviant for silver extraction.
With thiosulphates	Important complexes in photography, depending on concentration of $S_2O_3{}^{2-}$ we can find the complex $Na_3\,[Ag(S_2O_3)_2]$ after the development of the image.
With thiourea	Thiourea $CS(NH_2)_2$ can be used as a silver lixiviant in an acidic environment.
With heterocyclic organic ligands with donor atoms like nitrogen or sulfur	Pyridine complexes, thiazole complexes.

1.3 Speaking Nano

In the technical-scientific field, nano- (abbreviated by the symbol n) is one of the prefixes originally established by the International System of Units (SI). Nano- is used for small fractions of the unit, in fact it means one billionth, in other words 1 nm corresponds to 10^{-9} meters (Table 1.8).

Nanoparticles are small, very much smaller than bulk material, but bigger with respect to single atoms that ranges from about 0.1 nm to 0.5 nm or molecules which are formed by atoms bonded together.

Since ancient times, finely ground metal powders were introduced into glass to realize colored jewels, decorations, and stained glass in churches. However, it was only in 19th century that the role of metallic nanoparticles on the optical properties was studied, by the eminent English scientist Michael Faraday (1791–1867) [57].

Table 1.8 Common prefixes by SI

Prefix	Symbol	Factor	Origin of the name
deci-	d	10^{-1}	Latin *decimus*: tenth
centi-	c	10^{-2}	Latin *centum*: hundred
milli-	m	10^{-3}	Latin *mille*: thousand
micro-	μ	10^{-6}	Greek *micros*: little, small
nano-	n	10^{-9}	Greek *nanos*, Latin *nanus*: dwarf
pico-	p	10^{-12}	Spanish *pico:*, Italian *piccolo*: small
femto-	f	10^{-15}	Danish or Norwegian *femten*: fifteen
atto-	a	10^{-18}	Danish or Norwegian *atten*: eighteen
zepto-	z	10^{-21}	Greek *epta*: seven
yocto-	y	10^{-24}	Greek *okto*: eight

The particular optical properties of metal nanoparticles are due to the interaction between the surface electrons and the incident light radiation. Depending on metal nanoparticles size, shape, and agglomeration state and on the nature and composition of the matrix in which nanoparticles are dispersed, a certain wavelength radiation can induce a collective oscillation of the conduction electrons originating the so-called surface plasmon resonance (SPR) effect. This is possible because the wavelengths of the visible light spectrum ranging from 400 nm to 750 nm are longer than the sizes of nanoparticles. Resonance peak of nanosphere-shaped noble metals of approx. 100 nm diameter appears in the visible absorption spectrum at 440 nm (blue) for silver and 520 nm (green) for gold. Therefore, when these silver and gold nanoparticles are introduced into a transparent medium, such as water, the colors yellow and red, respectively, are obtained.

Optics is just one aspect affected by nanosize. In general, the physical properties (optical, magnetic, conductivity, melting point, etc.) and chemical properties (reactivity, stability, toxicity, heat of combustion, etc.) of objects below 100 nm can be drastically different from those of larger objects that follow classical physics and quantum chemistry.

The dependence of the properties of matter on its dimensions is one of the main reasons why nanomaterials have such a high potential. In fact, resizing to the nanoscale a certain material can completely change its properties and open new possibilities. The capability to control the size of particles is extremely important in order to obtain the desired characteristics, and the true challenge is to observe our new properties.

1.3.1 Observation

The microscope (from Greek μικρόν *mikrón*, which means little and σκοπεῖν *skopéin*, which means to watch) is the instrument that allowed to investigate first in the field of microscale, and then in nanoscale with particular modifications.

The observation of things trying to see the smallest details dates to the Romans (c. 1st century) who through the glass were able to see the larger objects. The first glasses date back to the Florentine inventor Salvino D'Amato (1258–1312) in 13th century. During the 1590s two spectacle makers Zacharias Jansen (1580–1632) and his father experimented with the first microscope consisting of two lenses mounted in a tube. Finally, in 1590 the Italian scientist Galileo Galilei (1564–1642) developed a two-lenses microscope, a concave and the other convex. A schematic timeline about the development of microscope is shown in Fig. 1.12.

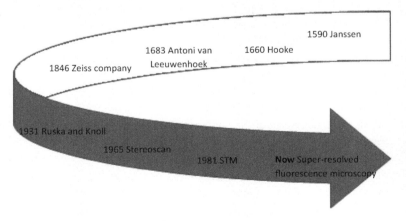

Figure 1.12 Microscope timeline.

The microscope was widespread since the 17th century by the English scientist Robert Hooke (1635–1703) and Dutch scientist Antoni van Leeuwenhoek (1632–1723).

All the scientific works of Antony van Leeuwenhoek about natural sciences are obtained by observations with a simple spherical lens devised by him. Leeuwenhoek discovered and studied many microbes, and bacteria.

The first work based on use of composed[6] microscope is *Micrographia* of Hooke, published in London in 1665. The instrument built by Hooke was composed by three biconvex lenses and by a spherical bowl filled with water. This bowl pitcher was necessary to focus the beam on the object. The light illuminates the object from above.

During the seventies of the 19th century, German physicist Ernst Abbe (1840–1905) developed a physical theory of the microscope that was applied by the company founded in 1846 by Carl Zeiss. This company will become famous all over the world to produce scientific instruments. With Abbe began the collaboration between physics, biomedical research, and manufacturers very important to develop new scientific instruments.

The contribution of Abbe to the theory of microscope as well as the extraordinary skill of craftsmen of Zeiss ensured great success to the company, which in 1880 already employed some 80 workers. At that time, further improvements were made to the microscope thanks to the collaboration of the German chemist Otto Schott (1851–1935), who conceived a new range of optical glasses that allowed a better correction of aberrations. In 1904, as a direct consequence of the Abbe formula, Zeiss produced a microscope in which visible light was replaced by ultraviolet light. The lower wavelength allowed to further increase the resolution power of the microscope.

In 1931, two German engineers of TU (Technische Universität) of Berlin Ernst Ruska (1906–1988) and Max Knoll (1897–1969), obtained the first electronic enlarged of an object. Electronic microscopes seemed less promising to study biological samples because the sample had to be in vacuum and dehydrated. The electronic microscope seemed more apt to study materials science rather than biology or medicine. In 1937 Ruska managed to

---

[6]Optical instrument that allows you to see more details than those obtained with a simple magnifying glass.

convince the company Siemens to support his research. A year later the company start a new laboratory managed by the brother of Ruska, the physician Helmut Ruska (1908–1973), the purpose of the laboratory was to demonstrate the potential applications to biology and medicine.

In 1939, when Siemens produced the first commercial microscope, the laboratory had published more than twenty scientific papers about biological structures clear only by electronic microscope.

The first scanning electron microscope (SEM) was built thanks to drive of Sir Charles Oatley (1904–1996) of University of Cambridge (UK). The project started in 1948, and in 1951 the first instrument was used. However, the first images of high quality were obtained in 1960. The first commercial SEM named Stereoscan, was assembled in 1965 Cambridge Instrument Company.

In 1986 the Nobel Prize in Physics was divided between, Ernst Ruska "for his fundamental work in electron optics, and for the design of the first electron microscope," Gerd Binnig (1947–now), and Heinrich Rohrer (1933–2013) "for their design of the scanning tunneling microscope."

Thanks to this and other tools like XRD (X-ray diffraction), BET (Brunauer–Emmett–Teller method), IR (infrared) and Raman spectroscopy, and DLS (dynamic light scattering), it was possible investigate about many properties of nanostructured materials.

1.3.2   Size Effect in Nanoparticles

Nanoparticles are structures (molecular or atomic) whose size is equal or inferior to 100 nm, when nanoparticles are aggregate like a grape we can talk about nanoclusters.

The nanoclusters can only be formed from a certain number of particles, this number of particles is called *magic number*; magic number shows how clusters are formed from single particles. The first magic number is 13, which represents the packing of 12 identical spheres the surrounding an internal sphere. The number of particles N_n, in the nth shell will be calculated according to the formula:

$$N_n = 10\, n^2 + 2 \qquad (1.16)$$

Therefore, if in the first shell there are 12 spheres, in the second there are 42 spheres; the sum of two shell give 53, the subsequent magic number. Following magic numbers will be 55, 147, 309, 567, and so on.

When we reduce the size of nanometric materials, we influence structural, chemical, thermodynamic, electron, spectroscopic, and electromagnetic properties. It is a particularly interesting aspect for scientists because the size control allows to modify the properties of the material.

We mainly talk about two types of effects that depend on the size: the different number of surface atoms and the quantum effect. We will also see what happens when electromagnetic radiation interacts with nanoparticles.

1.3.2.1 Surface atoms

A nanometric system, containing only a few hundred atoms, is largely influenced by the fact that most of these are on the surface, in a state of greater internal energy.

We may consider a cluster constituted of N atoms, whose form is close to spherical.

Its volume is defined by

$$V = \frac{4}{3}\pi r^3 = V_0 N \qquad (1.17)$$

where R is the radius of a sphere and V_0 is the volume that corresponds to a single atom.

We may define also V_0 of the sphere of radius a like

$$V = \frac{4}{3}\pi a \qquad (1.18)$$

While the area of the sphere is

$$A = 4\pi a^2 \qquad (1.19)$$

Nanoparticles have a high area/volume ratio, which means that this magnitude depends on a^{-1}; the number N of atoms contained in this sphere depends linearly on the volume, so with a^3, so the area/volume ratio will also depend on $N^{-1/3}$.

In a particle, the atoms on the surface have fewer atomic neighbors than the bulk ones, atoms near the corner, edge, and outer surfaces on the plane have a low cohesive energy, and greater affinity for forming bonds with adsorbed molecules.

The equation of Gibbs–Thomson [Eq. (1.20)], shows how the melting temperature of a group of atoms, cluster, depends on the inverse of the radius:

$$\frac{T_m - T_{mb}}{T_{mb}} = \frac{2V_m(l)\,\gamma_{sl}}{\Delta H_{mr}} \tag{1.20}$$

In this equation, T_m represents the melting point of the radius cluster r, T_{mb} the melting point of the bulk, V_m (l) the molar volume of the liquid, γ_{sl} the interfacial tension between the surface of the solid and that of the liquid and ΔH_{mr} the latent heat of fusion.

The surface energy of the liquid phase is always lower than that of the solid phase since the fluid phase is dynamic and allows the atoms to move to minimize the surface area and unfavorable interactions while in the solid the atoms are constrained by rigid geometry of ties, in high energy situations. By fusion, total surface energy decreases, and as noted, the decrease in size anticipates the fusion process because it increases the contribution of superficial energy. Surface effects are important whether they are nanoparticles in the form of dust or dispersed in a solvent (colloids), or in the form of a thin film with a nanometric structure.

It has been experimentally found that for the silver bulk the melting temperature is 961.78°C, for nanoparticles with a diameter of 1.5 nm it goes down to about 770°C [58].

Increasing the number of surface atoms also increases chemical reactivity; for example, in a 1 nm crystal, almost all the atoms are on the surface, while increasing the size of an order of magnitude we will only have about 15% of the atoms on the surface. This is very important because any interaction of nanoparticles with the external environment occurs through the atoms to the surface.

For example, if the nanoparticle is used as a sensor for environmental monitoring, the sensor will be much more effective [59]. When nanoparticles change shape passing from nanosphere to nanoprisms and nanowires we observe a different reactivity, an example is the antibacterial effect [60].

1.3.2.2 Quantum effect

The electronic structure of the nanoparticles varies as their size decreases, yet.

We may define an atomic orbital like a wave function that describes the probability that electrons of an atom are in a certain area around the nucleus. The interaction of several atoms produces an overlapping of atomic orbitals, this overlapping gives rise to molecular orbitals with energies that so are similar to each other to generate a continuous band of energy.

In the band theory for solid, N atoms with N atomic orbitals form N molecular orbitals (MO) [61]. Each MO can have two opposed-spin electrons so that $2N$ energy states will occur. These were discrete but for large values of N the space between them is so small that it can be considered continuous so as to form an energy band.

We can explain DOS (density of states) like the number of energy states available per unit of energy per unit of volume; they are used as units $J^{-1}m^{-3}$ or $eV^{-1}cm^{-3}$. DOS gives information on how energy states are distributed in a solid data; typically, it is indicated with $g(E)$ and can be measured by techniques such as STM or EELS (electron energy loss spectroscopy). For a high DOS system at a specific energy there will be many occupational energy levels, whereas if the DOS is zero there will be no occupational energy.

In a metal or semiconductor, the valence band is the highest energy band which, at sufficiently low temperatures, is completely filled with electrons. The electrons that occupy this band correspond to the valence electrons of the atoms that make up the solid. Some of the higher energy levels of the valence band can become free following excitation of the electrons that occupy them in the higher levels, or the presence, in the solid grid, of appropriate impurities; in this case the remaining electrons can be redistributed on the levels within the valence band, for example under the action of an electric field, and this from a place to electric current.

The conduction band is defined as the lowest energy band with available levels. This is separated from the valence band by an energy range called E_g energy gap in which electrons cannot be. In the metals, the two bands are superimposed, in the semiconductors there will be a certain energy gap, greater if the material is insulating.

When in a metal the valence electrons pass to higher levels they are called conduction electrons and generate electric current.

The separation between the filled states and the empty states is called the *Fermi energy* and is denoted by E_F. Another way of defining the Fermi energy is that this is the highest filled state in a metal at 0 K. The energy required to remove an electron from the Fermi level to vacuum level (where the electron is free of the influence of the metal) is called the *work function*. Energy levels can be referenced either with respect to the vacuum level (takes as zero) or with respect to the bottom of the valence band. Thus, metals like Li have a partially filled valence band.

The energy gap δ between two electronic levels can be defined by the following equation:

$$\delta = \frac{4\,Ef}{3n} \tag{1.21}$$

in which E_F is the Fermi energy level of the bulk material, and n is the total number of valence electrons in the nanoparticle.

Fermi level is the term used to describe the top of the collection of electron energy levels at absolute zero temperature. This concept comes from Fermi–Dirac statistics. The *fermions*, named in honor of the Italian physicist Nobel laureate Enrico Fermi, have spin, or angular momentum, half-integer body and following the principle of Pauli exclusion cannot exist on the same energetic state. Therefore, at absolute zero they pack into the lowest available energy states and build up a *Fermi sea* of electron energy states. The Fermi level is the surface of that sea at absolute zero where no electrons will have enough energy to rise above the surface. The idea of the Fermi energy is very important to understand the electrical and thermal properties of solids. Both ordinary electrical and thermal processes involve energies of a small fraction of an electron volt, the Fermi energies of metals are approximately electron volts. This involves that the large number of the electrons cannot receive energy from those processes because there are no available energy states for them to go to within a fraction of an electron volt of their present energy. Limited to a tiny depth of energy, these interactions are limited to "ripples on the Fermi sea."

At higher temperatures a certain fraction, characterized by the Fermi function, will exist above the Fermi level. The Fermi level plays an important role in the band theory of solids. In doped semiconductors, p-type and n-type, the impurities present shift the Fermi level. In metals, the Fermi energy gives us information about the velocities of the electrons that participate in ordinary electrical conduction. The amount of energy which can be given to an electron in such conduction processes is on the order of micro-electron volts (see copper wire example), so only those electrons very close to the Fermi energy can participate.

The so-called *quantum size effect* describes the physics of electron properties in solids with great reductions in particle size. Quantum effects can begin to dominate the behavior of matter at the nanoscale impacting the optical, electrical, and magnetic behavior of materials.

The electronic structure of nanomaterials is different from its bulk material, the density of the energy states in the conduction band changes. The ionization potential is higher than that for the bulk material; the magnetic moment of nanoparticles is found to be very less when compared them with its bulk size. Finally, nanoparticles made of semiconducting materials germanium, silicon, and cadmium are not semiconductors.

For a bulk metal, the number of delocalized electrons in the band structure is equal to the number of atoms in the material mass. Under a normal temperature, the metal delocalized electrons can be easily promoted to a state of higher energy, and can move freely in the structure. This gives the material its nature electrically conductive. In traditional semiconductor materials, the number of delocalized electrons is significantly lower than the number of atoms. This is highlighted in a higher value of E_g which is significant at ambient temperature. This means that in a semiconductor the electrons will not be free to move and conduct current, without a certain additional energy input.

For nanoparticles, the electronic energy levels are not continuous unlike the bulk but discrete. The reason is the confinement of the wave function because of the physical dimensions of the particles.

Other examples, which involve more extensive nanometric structures, are the catalytic activity of metal particles of 1–2 nm (e.g., gold), the hardness of carbon from diamond to graphite is very different,[7] and the color of a crystal of CdS is very diverse when the same is reduced to few nanometers [62].

For nanocrystals of different metals, the gap between the conduction and the valence band is to be the one typical of the non-metals [63].

It is possible to apply Eq. (1.21) to a nanoparticle of Ag of 3 nm diameter and about 1000 atoms (and then ~1000 valence electrons) would have a δ value of about 5–10 meV. If the thermal energy, kT, is greater than the Kubo gap the nanoparticle would be of metallic nature, but if kT fall below the Kubo gap it would become non-metallic [64]. At room temperature kT is about 26 meV, and then a silver nanoparticle of 3 nm would exhibit metallic properties. If the size of the nanoparticle was decreased or the temperature was lowered, the nanoparticle would show a non-metallic behavior [65].

Using this theory, with a Fermi energy value for the bulk silver equal to 5.5 eV, then the silver nanoparticles would cease to be metallic when below of about 280 atoms, at room temperature. For this reason, in the nanoparticles, properties such as electrical and magnetic susceptibility of conductance show quantum size effects. These effects have led the nanoparticles to be used in many applications from catalysis, optics up to medicine. These effects have led the nanoparticles to be used in many applications from catalysis, optics up to medicine.

Thus, with decreasing size, the effective bandgap increases, and the relevant absorption and emission spectra shift to bluer wavelengths.

Furthermore, the qualitative variation of the electronic structure due to the confinement, for small nanocrystals, confers unusual catalytic properties to these particles, with different behaviors from bulk. For example, from studies on the interaction of silver nanocrystals of various sizes and molecular O_2, it is observed that the metal nanoparticles transform molecular oxygen into atomic oxygen, this phenomenon is not appreciated for the material in bulk. The catalytic behavior is inferred from the change of binding energy of silver, exposed to a flow of oxygen (500L at 80 K). At the massive metal it binds to the molecule of O_2, while the nanocrystals interact with the single atom, O^- [66].

---

[7]Hardness of diamond 10 Mohs; graphite 1.5 Mohs.

1.3.2.3 Color

Sir Isaac Newton (1642–1727) was the first to carry out scientific experiments on the dispersion of light, famous to show that color is not a property of bodies but is due to a property of light. Using a glass Newton was able to break the sunlight into the so-called spectrum of the iris, all the colors were already present in the "white" light of the sun before decomposing it (Fig. 1.13).

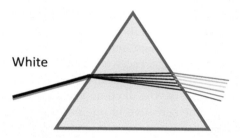

Figure 1.13   Light dispersion in a prism.

To confirm his hypothesis Newton recomposed the colors using a large converging lens, obtaining a white ray of light again. Newton distinguished seven colors, among which also indigo, between violet and blue, at the time of Newton the number seven was considered a perfect number.

Newton drew the circle of colors on which the colors of the spectrum were reported in sectors whose width was in relation to that observed in the spectrum. The position of the colors on the circle defined the quality relationships between the colors themselves, Newton imagined that between the colors there could be harmonic relations as between the seven musical notes, and that the colors close to each other (adjacent) develop harmonic relationships, while the colors that were in opposition (complementary) had a dynamic relationship between them. In Fig. 1.14, we can see the picture of correspondence between the different colors.

We must get to the Dutch scientist Christiaan Huygens (1596–1687) to formulate the wave theory of light (1678), according to which from every point of a light source they originate longitudinal spherical waves. The scientist Thomas Young (1773–1829) in the experiment of the double slit interference of 1801 passed a light beam through two parallel slots made on an opaque screen, in order to obtain a scheme of light and dark bands on one white surface behind the screen. This convinced Young of the wave nature of light. He described the phenomenon of astigmatism of the eye and the perception of colors. In reality it is the eye that distinguishes the different colors. The human eye can capture more than two hundred shades of color, and if each of these shades requires a type of photoreceptor, there should be more than two hundred different types of photoreceptor on the retina, one for each color, which is impossible considering the smallness of the retina surface. The photoreceptors are nerve endings connected to the brain, present on the retina of the eye and sensitive to light. Thanks to these photoreceptors, the image received from the eye is transmitted to the brain. Thomas Young advanced the hypothesis that the receptors responsible for the daytime vision (color vision) were of three types, each one sensitive to one of the three primary colors that he established to be green, red, and blue. His hypothesis was only later confirmed by subsequent research on the physiology of the eye, which showed the existence on the retina of three different types of cones (photoreceptors used by the eye in daylight). Young hypothesized that the vision of the different shades of colors was generated by the combined action of these three receptors, able to react differently to the different frequencies present in the light radiation. In this work of color perception, the brain plays a fundamental role, that of taking the information from the eye and returning it to our perception in the form of a colored visual image, as he had intuited in 1600 Descartes.

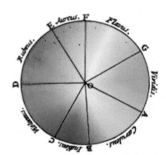

Figure 1.14 Newton's color circle.

Light scattering is part of our daily experience. The color of the sunset, the blue sky, and the gray clouds are all manifestations of the scattering of light. We can define Rayleigh scattering as a phenomenon in which the particles, suspended in a transparent medium, liquid or gas, are smaller than the wavelength of the incident light. Let's talk about Mie scattering when the case does not depend on the particle size. In the latter case, the method will have a more complex solution than Rayleigh's approximation. In both cases we speak of elastic scattering, when the frequency of the radiation associated with the process does not change. Examples of inelastic scattering, that is, when a change in the frequency of radiation is observed, is associated with two types of spectroscopic techniques called Brillouin and Raman.

Rayleigh equation defines the intensity of the scattered light:

$$I = I_0 \frac{8\pi N\alpha 2}{\lambda 4r2}(1 + \cos 2\theta) \qquad (1.22)$$

where I is the intensity of the scattered light, I_0 is the intensity of the incident light, α is the polarizability, r is the particle distance, N is the number of scattering particles, λ is the wavelength, and θ is the angle of incident light.

The intensity of the scattered light is inversely proportional to the wavelength of the incident radiation. The blue, with λ equal to 460–480 nm is then diffused with a much higher intensity than red, with λ equal to 620–700 nm. This explains the blue color of the sky.

To explain the different color of the metallic nanoparticles according to their size, we must introduce the concept of *surface plasmon*. In a metal the electrons can move freely, when it comes to nanoparticles the ability of electrons to move is limited to the metal surface. The collective oscillation of the electrons is called the surface plasmon. Plasmonic resonance is a coherent oscillation of surface plasmons excited by incident electromagnetic radiation. Plasmonic resonance determines the absorption of light and therefore the color of metallic nanoparticle solutions [67].

The SPR frequency of nano-sized metal particles is different from that of bulk material and has been shown to strongly depend on their size, shape, aggregation, and structure (solid vs. hollow), as well as the dielectric properties of surrounding media. For instance, the SPR peak of a thin film of gold is located at ~480 nm, while the SPR peak of an aqueous dispersion of 13 nm spherical gold colloids is centered around 520 nm. We might consider silver or gold nanoparticles as inorganic chromophores with strong extinction.

Depending on the shape and size of the nanoparticles we will have different colors. In the case of silver based on size and shape we will have colors ranging from yellow to green. The maximum molar extinction coefficient for spheres is around 400 nm, for triangular plates around 700 nm for silver disk around 600 nm [68].

1.4 Today's Application of Silver Nanoparticles

Nanotechnologies can play a key role in the value chain of a wide range of products and processes, enabling new components, systems, and processes with improved or totally new performances, effectiveness, functionality.

The use of silver nanoparticles is very frequent, albeit lower than other nanomaterials such as carbon black or silica. Many products are often of daily use. For this reason, there is an increasing focus on silver nanoparticles. The nanotoxicity of silver becomes very important at this point. It is necessary to characterize the silver nanoparticles-based materials as much as possible to understand their potential impact. Very recently, specific regulatory actions for nanomaterials have been considered regarding cosmetics, foods, biocidal products, while other are under discussion, in particular regarding chemicals.[8]

As for any other hazardous substance, risk is determined by the combination of hazard and exposure (quantity and way of dispersion into the different environmental media). Several data are available regarding the presence of silver derived from human use in the environment, and specific threshold level are set by environmental regulation. The knowledge and experience on silver, as a substance itself, and on the recent application of silver as a biocidal product, provides a relevant background in the evaluation of risks and benefits of silver nanoparticles. Unfortunately, references to products with related characterizations change rapidly due to corporate changes, geographical location, and classification methods.

---

[8]http://ec.europa.eu/research/industrial_technologies/nanoscience-and-technologies_en.html

It is important to evaluate the mobility in the human body and, in principle, in other biological media. Silver nanoparticles could be able to distribute in almost all human organs, as well as to penetrate cell membranes. The increased bioavailability could change the potential impact on human health and the environment, both in terms of hazard and exposure in the different biological media.

There are several databases representatives of number, and typology of products based of silver nanoparticles. Patents analysis shows that there are very many materials based on silver nanoparticles, especially in medical and health care, cosmetics, and clothing. Top patenting countries include USA, Korea, and China (accounting together for more than 80% of total patents), followed by Germany and other countries like RU, EP, JP, GB, TW, UA, IT, IN, FR, DK, ES.

That of the household equipment is one of the largest field of application of silver nanoparticles. In products such as refrigerators, hair dryer and steam irons, silver nanoparticles seem to be used as a coating, providing an antibacterial effect by releasing silver ions during use. In filtration systems, such as in vacuum cleaners and air conditioners, silver nanoparticles seem to be used in the form of both coatings and nanoparticles, somehow embedded in the materials of the product (a vacuum cleaner producer is claiming to have "particles of antibacterial nano silver in dust bucket").

One of the most famous companies has produced a washing machine, using a silver plate that is slowly electrolysed to release silver ions during wash and rinse cycles to "sanitize" clothing. The product started a long debate in the USA, due to the request of the U.S. Environmental Protection Agency (EPA) to consider and regulate it as a pesticide (silver ions might concentrate in wastewater treatment plants and kill bacteria used to detoxify the wastewater).

Silver nanoparticles may be incorporated in clothing and textiles during different steps of the manufacturing process, like a coating or before the extrusion of fibers. The purpose is to avoid the formation of bad smells and molds.

Antimicrobial fabrics-based silver nanoparticles can have applications in the medical textiles sector (hygiene textiles) and health-related fields, such for hospital settings (linen, curtains, and other furniture that needs antibacterial properties), or for bandage and wound dressings. In the latter case, both fabric or other type of supports can be used to store silver ions and ensure a time-release mechanism to act against bacteria. Other uses in the medical field includes antibacterial coatings for implants, catheters, and others.

There are also cosmetic based on silver formulations, especially to treat bacterial, fungal, and viral infection.

Many oxidation reactions use silver nanoparticles as a catalyst, for example, in the production of formaldehyde from methanol [69] or in the catalytic reduction of NO_x to N_2 with alumina [70].

Ink-jet printing is a low technique to fabricate electro-circuits for direct writing of patterns, instead of photolithography. Printed inks contain metallic nanoparticles, carbon nanotubes, and ceramic nanoparticles. A major challenge for depositing materials is the formulation of suitable ink; ink formulation determines the drop ejection characteristics and the quality of the printed electro-circuits. Viscosity, to a certain extent, determines the formation of drops and, the break-up and corresponding tail and surface tension is responsible for the spherical shape of the liquid drops emerging from a nozzle. It is possible to jet a wide variety of fluids with viscosities in the range of 2–30 mPa s and surface tensions up to 60 mN m^{-1} [71].

Silver nanoparticles have a good conductivity, and there are many methods of synthesis.

In the field of conductive polymers, silver nanowires-filled methylcellulose has been produced [72]. These composites show good electrical properties and enhanced mechanical properties.

Chinese scientists have tested a cholesterol biosensor based on enzymatic silver deposition [73], the dispositive has shown very good response to variation of cholesterol concentration.

The phenomenon of photocatalytic splitting of water on titanium oxide electrode under ultraviolet light has been known for many years. Silver nanoparticles deposited on the surface of titanium oxide improve photocatalysis. The purpose is to degrade organic pollutants [74].

These are just some of the possible applications of nanoparticles with other components.

References

1. Barney, S.A., Lewis, W.J., Beach, J.A., and Berghof, O. (2006) *The Etymologies of Isidore of Seville*, Cambridge University Press.

2. Mahdihassan, S. (1988) *Am. J. Chin. Med* **16**, p. 83.

3. Mahdihassan, S. (1984) *Anc. Sci. Life* **IV** (2), pp. 116–122.

4. Ball, D.W. (1985) *J. Chem. Educ.*, **62** (9), pp. 787–788.

5. Rasmussen, S.C. (2015) *Chemical Technology in Antiquity*, ACS Symposium Series, Vol. 1211 Chapter 5, pp. 139–179.

6. Holy Bible (23.1 to 23.20).

7. (a) Cobb, C., Fetterolf, M.L., and Goldwhite, H. (2014) *The Chemistry of Alchemy*, Prometheus Books, NY. (b) Sneed, M.C., Maynard, J.L., and Brasted, R.C. (1954) *Comprehensive Inorganic Chemistry* Vol. II, p. 114, D. Van Nostrand Company Inc., NY.

8. Bunker, E.C. (1994) *Orientations* **25**, pp. 73–78.

9. Hörz, G., and Kallfass, M. (1998) *The Journal of The Minerals, Metals & Materials Society* **50**, pp. 8–16.

10. Rehren, T. (2009–2011) *Archeology international* **13/14**, pp. 76–83.

11. Cesareo, R., Calza, C., DosAnjos, M., Lopesb, R.T., Bustamante, A., Fabian, J.S., Alva, W., and Chero, L.Z. (2010) *Applied Radiation and Isotopes* **68**, pp. 525–552.

12. (a) Schwartz, P. (2017) *Alchemist* **87**, pp. 3–5. (b) Haring, C.H. (1915) *The Quarterly Journal of Economics* **29** (3), pp. 433–479.

13. Ercolani Cocchi E. (2003) *Dal Baratto all'euro*, Editoriale Olimpia.

14. Pense, A.W. (1992) *Materials Characterization* **29**, pp. 213–222.

15. Redish, A. (2006) *BimetallismAn Economic and Historical Analysis* (Studies in Macroeconomic History), Cambridge University Press.

16. Deng, K.G. (2008) *Pacific Economic Review* **13** (3), pp. 320–358.

17. Polo, M. and Da Pisa, R. (2010) *Il Milione*, Classics Publishing.

18. Sir Alexander Cunningham (1891) *Coins of Ancient India: From the Earliest Times Down to the Seventh Century A.D.*, Asian Educational Services.

19. Cribb, J. (2003) *South Asian Studies* **19** (1), pp. 1–19.

20. Cox, R.J. (1973) *Photographic Processing*, Academic Press.

21. (2008) *Annals of the ICRP* **37**, pp. 2–4.

22. (2009) Bollettino dei Musei Civici Veneziani III Serie. Le medaglie rinascimentali di scuola veneziana nelle collezioni dei musei civici veneziani. Marsilio, Venice, Italy.

23. Camic, B.T., Oytun, F., Aslan, M.H., et al. (2017) *Journal of Colloid and Interface Science* **505**, pp. 79–86.

24. Steinbrück, M., Stegmaier, U., and Grosse, M. (2017) *Annals of Nuclear Energy* **101**, pp. 347–358.

25. Zhang, X., Ren, W., Zheng, Z., and Wang, S. (2019) *IEEE Access* **7**, pp. 133079–133089.

26. Joshi, D. (2011) *Rasa Sastra* English edition, Chaukhambha Orientalia.

27. Barve, G.M., Mashru, M., Jagtap, C., Patgiri, B. J., Prajapati, P.K. (2011) *Journal of Ayurveda & Integrative Medicine* **2** (2), pp. 55–63.

28. Martin del Barco Centenera *La Argentina La conquista del Rio de La Plata. Poema historico Argentina y conquista del Rio de la Plata, con otros acaecimientos de los Reinos del Perú, Tucumán y Estado del Brasil* (Spanish Edition), University of Michigan Library.

29. Giovannelli, G., Natali, S., Bozzini, B., Manno, D., Micocci, G., Serra, A., Sarcinelli, G., Siciliano, A., and Vitale, R. (2006) *A puzzling Mule Coin from the Parabita Hoard: A Material Characterisation*, Cornell University.

30. La Niece, S., Harris, V., and Uchida, H., (2014) *ISIJ International* **54** (5), pp. 1111–1116.

31. (a) Giumlia-Mair, A. (2000) Argento e leghe "argentee" nell'antichità, 7° Convegno "Le Scienze della Terra e l'Archeometria", Bollettino Accademia Gioenia di Scienze Naturali **33** (357), pp. 295–314. (b) Giumlia-Mair, A. (2008) *Surf. Eng.* **24** (2), pp. 110–117.

32. Weast, R.C. (1984) *CRC Handbook of Chemistry and Physics* 64th Edition, B258–B259, CRC Press, Boca Raton, Florida.

33. Schönbächler, M., Carlson, R.W., Horan, M.F., and Hauri, E.H. (2010) *Science* **328**, pp. 884–887.

34. Webster, A.J., and Mann, G.W. (1984) *Journal of Geochemical Exploration* **22** (1–3), pp. 21–42.

35. www.silverinstitute.org

36. Guerra, M.F., and Abollivier, P. (2016) *Nucl. Instr. Meth. Phys. Res. B* **377**, pp. 1–11.

37. Agricola, G. (1556) *De re metallica.*

38. Benner, R.C., and Hartmann, M.L. (1911) *J. Ind. Eng. Chem.*, pp. 805–807.

39. Nriagu, J.O. (1985) *J. Chem. Educ.* **62** (8), pp. 668–674.

40. Dorr, J.V.N, and Bosqui, F.L. (1950) *Cyanidation and Concentration of Gold and Silver Ores* 2nd Edition, pp. 428–445, McGraw-Hill Book Co., Inc., New York.

41. Syed, S. (2016) *Waste Management* **50**, pp. 234–256.

42. Gurung, M., Adhikari, B.B., Kawakita, H., Ohto, K., Inoue, K., and Alam, S. (2013) *Hydrometallurgy* **133**, pp. 84–93.

43. Aktas, S. (2010) *Hydrometallurgy* **104**, pp. 106–111.

44. Valverde, F., Costas, M., Pena, F., Lavilla, I., and Bendicho, C. (2008) *Chemical Speciation & Bioavailability* **20** (4), pp. 217–226.

45. McMillan, J.A. (1962) *Chem. Rev.* **62** (1), pp. 65–80.

46. Patent US2004022868 (A1) Antelmann, M. *Compositions using tetrasilver tetroxide and methods for management of skin conditions using same.*

47. Pan, J., Sun, Y., Wang, Z., Wan, P., Liu, X., and Fan, M. (2007) *J. Mater. Chem.* **17**, pp. 4820–4825.

48. Hammer, R.N., and Kleinberg, J. (1953) *Inorg. Synth.* **1** (4), p. 12.

49. Waterhouse, G.I.N., Bowmaker, G.A., and Metson, J.B. (2002) *Surf. Interface Anal.* **33**, p. 401.

50. Patent WO2008120259 (A1) Bogana, M.P., Bottani, C.E., Dellasega, D., Di Fonzo, F., and Facibeni A. *Material of nano-aggregates of tetrasilver tetroxide.*

51. Sadovnikov, S.I, Gusev A.I., and Rempel A.A. (2015) *Superlattices and Microstructures* **83**, pp. 35–47.

52. (a) Frencken, J.E., Peters, M.C., Manton, D.J., Leal, S.C., Gordan, V.V., and Eden, E. (2012) *Int. Dent. J.* **62**, pp. 223–243. (b) Butt, N., Talwar, S., Chaudhry, S., and Nawal R.R. (2013) *Indian J. Dent. Adv.* **5**(2), pp. 1186–1194.

53. Configliachi, P., and Brugnatelli, G. (1824) *Giornale di Fisica, Chimica, Storia Naturale, Medicina, ed Arti.*

54. https://www.nobelprize.org/nobel_prizes/chemistry/laureates/1913/

55. Hartshorn, R.M., Hellwich, K.H., Yerin, A., Damhus, T., and Hutton, A.T. (2015) *Pure Appl. Chem.* **87** (9-10), pp. 1039–1049.

56. (a) Korany A.A., Mokhles, M. A-E., and Mahmoud, K. (2013) *International Journal of Medicinal Chemistry*, pp. 1–7. (b) Johnson N.A., Southerland M.R., and Youngs W.J. (2017) *Molecules* **22**, pp. 1263–1283.

57. Faraday, M. (1857) *Phil. Trans. R. Soc.* **147** pp. 145–181.

58. Little, S.A., Begou, T., Collins, R.W., and Marsillac, S., (2012) *Applied Physics Letters* **100**, pp. 051107–051110.

59. Rassaei, L., Amiri, M., Cirtiu, C.M., Sillanpää, M., Marken, F., and Sillanpää, M. (2011) *Trends in Analytical Chemistry* **30** (11), pp. 1704–1715.

60. Van Dong, P., Ha, C.H., Binh, L.T., and Kasbohm, J. (2012) *International Nano Letters* **2**, pp. 1–9.

61. Pauling, L. (1960) *The Nature of the Chemical Bond and the Structure of Molecules and Crystals: An Introduction to Modern Structural Chemistry*, Cornell University Press.

62. (a) Henglein, A. (1989) *Chem Rev.* **89**, pp. 1861–1873. (b) Miyake, M., Torimoto,T., Sakata, T., Mori, H., Kuwabata, S., and Yoneyama, H. (1997) *Langmuir* **13**, pp. 742–746.

63. Alivisatos, A.P. (1996) *J. Phys. Chem.* **100**, p. 13226.

64. Roduner, E. (2006) *Chem Soc. Rev.* **35**, pp. 583–592.

65. (a) Edwards, P.P., Johnston, R.L., and Rao, C.N.R. (1998) in *Metal Clusters in Chemistry* (Eds., P. Braunstein, G. Oro, P. R. Raithby), Wiley-VCH. (b) Kirkland, A.I., Jefferson, D.A., and Duff, D.G. (1993) Annual Reports C, *Royal Society of Chemistry*, p. 247.

66. Rao, C.N.R., Kulkarni, G.U., Thomas, P.J., and Edwards P.P. (2002) *Chem. Eur. J.* **8**, pp. 28–35.

67. Bohren, C.F., and Human, D.R. (1983) *Absorption and Scattering of Light by Small Particles*, Wiley, New York.

68. (a) Kreibig, U., and Genzel, L. (1985) *Surf. Sci.* **156**, pp. 678–700. (b) Yu, Y.Y., Chang, S.S., Lee, C.L., and Wang, C.R.C. (1997) *J. Phys. Chem. B* **101**, pp. 6661–6664. (c) Haynes, C.L., and Van Duyne, R.P. (2001) *J. Phys. Chem. B* **105**, pp. 5599–5611.

69. Wachs, I.E., and Madix, R.J. (1978) *Surf. Sci.* **76**, pp. 531–558.

70. Lindfors, L.E., Eranen, K., Klingstedt, F., and Murzin, D.Y. (2004) *Topics Catal.* **28**, pp. 185–189.

71. (a) Zhang, Z., Zhang, X., Xin, Z., Deng, M., Wen, Y., and Song, Y. (2011) *Nanotechnology* **22**, pp. 425601–425609. (b) Tai, Y.L., and Yang, Z.G. (2012) *Surf. Interf. Anal.* **44**, pp. 529–534. (c) Odom, S.A., Chayanupatkul, S., Blaiszik, B.J., Zhao, O., Jackson, A.C., Braun, P.V., Sottos, N.R., Scott, R.W., and Moore, J.S. (2012) *Adv. Mater.* **24**, pp. 2578–2581. (d) Kosmala, A., Wright, R., Zhang, Q., and Kirby, P. (2011) *Materials Chemistry and Physics* **129**, pp. 1075–1080.

72. Park, J.S., Ki, T., and Kim, W.S., (2017) *Scientific Reports*, pp. 1–10.

73. Huang, Y., Cui, L., Xue, Y., Zhang, S., Zhua, N., Liang, J., and Li, G. (2017) *Materials Science and Engineering C* **77**, pp. 1–8.

74. Maa, J., Yang, M., Sun, Y., Li, C., Li, Q., Gao, F., Yua, F., and Chen, J. (2014) *Physica E* **58**, pp. 24–29.

Chapter 2

Methods for Silver Nanoparticles Production

2.1 Build or Tear Down?

Approach: a way of dealing with a situation or problem. The problem to solve is the production of silver nanoparticles, and the question is how to produce by the less toxic, expensive, and suitable method.

We can see below the two approaches for the preparation of silver nanoparticles (see Fig. 2.1).

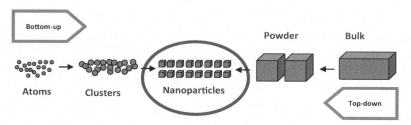

Figure 2.1   Scheme of the two-alternative approach.

We can start from a bulk material to reduce it to nanometric dimensions or start from atoms to grow nanoparticles. The kind of nanoparticles influences our choice, within this classification we can use physical or chemical methods

Silver Nanoparticles: Synthesis, Properties, and Applications
Anna Facibeni
Copyright © 2023 Jenny Stanford Publishing Pte. Ltd.
ISBN 978-981-4968-21-8 (Hardcover), 978-1-003-27895-5 (eBook)
www.jennystanford.com

2.1.1 Top-Down Approach

Top-down approach starts from a material bulk and break it into small pieces mainly by physical methods, but not only. For example, to evaporate a material from condensate phase you must provide energy or heat. If we use heat its saturated vapor pressure is very important, from this depend evaporation temperature.

We can do this through different processes such as thermal evaporation regulated by Joule effect or by sputtering, in which the material to be deposited is destroy by a plasma (powered DC or RF current) or spark discharge in which the evaporation is produced by electric discharge.

Pulsed laser deposition is another technique in which a high-power laser is focused through optics on a target that is placed in a vacuum chamber. Vaporized (or ablated) species form the plume,[9] and finally the material collects on a substrate that is present in the deposition chamber; very important is an evolution of pulsed laser deposition, the pulsed laser ablation in liquid, in which the ablated material is not ejected in one direction, but it is scattered inside the liquid.

The chemical top-down methods more used are dealloying or selective leaching, defined as the removal of one metal from solid solution of metals; or anisotropic dissolution, in which larger crystals are dissolved in a controlled way in under saturated solutions to create nanoscale features on the crystal surface.

2.1.2 Bottom-Up Approach

In this case we start from homogeneous solution or gas and build up the nanoparticle or nano-layer, nano-wire or whatever. Often makes use of deposition techniques (e.g., atomic layer deposition and chemical vapor deposition) and molecular self-assembly processes (e.g., self-assembled monolayers).

Chemical vapor deposition (CVD) and atomic layer deposition (ALD) [1] are based on the manufacture of materials from vapor phase. Both methods are not widely used in the production of biomaterials, although two methods allow exact control of the

---

[9]Migration of different ionized species from the target surface, thus generating a plasma of particles.

nucleation and growth of metallic single grains or layers. We will see the first one in Section 2.2.2.1.

The technique known by the acronym SAM (self-assembled monolayer) induces self-organization by exploiting the chemical bonds both with the substrate and intermolecular [2]. If organic molecules are deposited on a surface ("the substrate"), large-scale agglomerates can be formed spontaneously. The formation of these structures ("self-assembly" or "self-organization") is the result of the overall action of forces between the individual molecule and the substrate and of intermolecular forces.

The chain is generally formed by a succession of groups $-CH_2$. The interactions between the molecular "chains" determine the conditions of organization of the layer. A chemical group of "tail" still the substrate molecule. The "head" termination serves to modulate the chemical characteristics of the outermost interface, that is functionalize the surface.

Another extensively used bottom-up method is the reduction of ions in solution through rapid nucleation followed by controlled growth in the presence of coordinating ligands or surfactants. This technique represents a direct reaction to obtain silver nanoparticles, it can take place chemically, electrochemically or by radiolysis. Biological methods of silver nanoparticles synthesis are not simple, because those organisms are sensible to higher concentration of silver. This wet based technique produces nanocrystalline materials that can be further utilized as building blocks for the fabrication of functional nanoscale superstructures.

Through chemical reactions we can design nanoparticles size, shape, composition, and agglomeration. The control of this parameters is important to obtain properties as optical, magnetic, and electronic.

2.2 What to Choose Between Chemical and Physical Way?

At this point we can choose whether to follow physical or chemical pathway. The choice of method to produce AgNPs depends on our equipment, our background, and not our preferences. Generally, the production of nanoparticles by physical methods requires special equipment whether it be thermal evaporation, sputtering or PLD, and it could be more expensive compared to chemical methods.

Below a brief account of the main methods to produce silver nanoparticles.

2.2.1 Physical Methods

The main advantages of physical methods in comparison with chemical reaction processes are the absence of solvent contamination in the synthesized nanoparticles and the accurate control of their dimensions.

The physical vapor deposition (PVD) is an atomic deposition process in which the material is evaporated by a solid or liquid source in the form of atoms or molecules and transported into vapor in through an empty or plasma environment under the condensation. Different techniques have been employed to evaporate the source such as thermal evaporation, spark discharge, laser ablation.

We can identify three principal steps:

1. **Evaporation** in which the material is evaporated from the target.
2. **Carrying** in which the evaporated material is transported to the substrate to be coated.
3. **Condensation** in which the material nucleates and grows on the substrate forming the coating.

2.2.1.1 Thermal evaporation

The starting point of this technique is the Joule effect. The Joule effect is named after the British physician James Prescott Joule (1818–1889) in 1848 studying the nature of heat demonstrated the relationship between the current flowing in a resistance and the dissipated heat. These two quantities are proportional to each other. This experiment laid the foundations for the law of conservation of energy and the first law of thermodynamics, although it was only a qualitative measure. The true quantitative experiment was done seven years later by the same Joule [3], in which he obtained the most famous measurement by using the fall of graves. In this experiment the gravitational potential energy was transformed into heat which was measured by means of a calorimeter. This experiment made it possible to derive the mechanical equivalent of the calorie with good precision.

The material to be evaporated is placed in a metal crucible with a high melting temperature (e.g., tungsten or molybdenum) or quartz in which a high intensity current is passed (see Fig. 2.2).

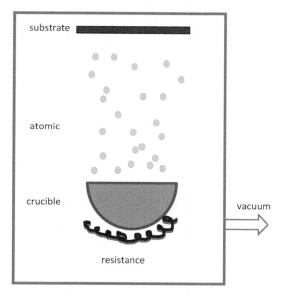

Figure 2.2 Scheme of thermal evaporation.

In this technique the deposition speed is high, it is not necessary to heat the substrate and it is an economical technique. Only materials with low evaporation voltage can be used as sources, therefore not ceramic materials, sodium, or potassium contamination may occur because they are present during crucible processing. In some cases, it is necessary to move the sample when there are steps or angled parts in the sample to be covered.

The thicknesses are limited by the capacity of the crucibles (or spirals) and are generally between 0.1 and 0.5 microns [4].

Thermal evaporation does not increase the temperature of the pieces to be coated and therefore is very suitable for deposition on plastic materials. Most applications of thermal evaporation with spirals concern the deposition of aluminum, in the automotive (e.g., for headlights), lighting, and decorative sectors.

2.2.1.2 Spark discharge

The electric arc, or arc discharge, is a phenomenon characterized by a very intense light emission that occurs between two electrodes. The electrodes (e.g., sticks of carbon or platinum or copper, etc.) are immersed in a gas (gas or steam). Their distance (from a few mm to a few cm) is varied so that an arc continuous is maintained. We give the name of anode to the electrode with greater potential, of cathode to the other.

The electric arc is characterized by a current density extremely high ($\gg$ 10 A/cm^2), the current near the cathode is essentially transported by electrons emitted by the cathode itself, and the potential difference between the electrodes is considerably lower than in other types of electrical discharge in the gases. The dazzling brightness that develops, a luminous arc that connects the ends of the electrodes, is accompanied by strong heat development.

The phenomenon was observed first by the English chemist H. Davy (1778–1829) in 1808. The technique was later developed by the American inventor T. H. Edison (1847–1931) [5].

Only the conductive materials can be vaporized, it is not possible to deposit oxides due to their high evaporation temperatures, the coating is irregular due to drops produced by the high current density [6].

To produce silver nanoparticles, we can use the electrodes in liquid, that is, distilled water, so we will have spark discharge in liquid (see Fig. 2.3).

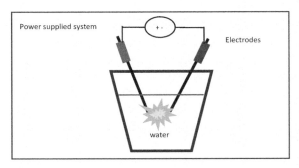

Figure 2.3 Spark discharge in liquid.

We can follow two different ways. The first is to produce the discharge between two silver electrodes in water the second one is to use titanium electrodes in an aqueous solution of silver nitrate. Tien et al. [7] reported a method to synthesize gold and silver nanoparticles by arc discharge without the use of stabilizers and surfactants in deionized water. The system consists in: silver electrodes (99.99%, 1 mm in diameter), a reactor for containing distilled water, a control system to maintain constant the gap between electrodes, and a power supply system to set the DC arc discharge parameters. In this case the initial voltage is 135 V, the intensity current 6.4 A, the pulse duration is 50s. During the arc discharge, both AgNPs and ions are produced. Tseng et al. [8] have used a method that consists of a combination of chemical methods and standard electrical discharge in liquid. They have added sodium citrate, the role of sodium citrate was to reduce the Ag ions into metallic silver, like mechanism of the Lee–Meisel method, and to cap silver nanoparticles increasing the zeta potential.[10] In this way smaller nanoparticles and a narrower size distribution were obtained. The resulting colloidal solution was extremely stable: after one year no precipitation occurred, and the zeta potential remained around its initial value. The conductivity of the solution increases with the concentration of sodium citrate. After a certain limit the plasma generated becomes unstable, leading to increasing in size of nanoparticles and less uniform size distribution.

2.2.1.3 Laser ablation deposition (LAD)

Laser is an acronym for **light amplification by stimulated emission of radiation**. This term indicates a device to obtain intense and extremely concentrated beams of coherent electromagnetic radiation in the infrared, visible, and ultraviolet fields.

In his paper of 1917 "On Quantum Theory of Radiation" the German scientist Albert Einstein (1879–1955) [9] presented the concept of amplification of radiation through stimulated emission with coherence. Only in 1952 Joseph Weber (1919–2000), developed the theory and started working on the construction of masers.[11] The first maser was demonstrated by Townes and his team in 1954 [10].

---

[10]Zeta potential is the potential generated following the formation of a double electric layer around the nanoparticles.
[11]Maser is acronym of microwave amplification by stimulated emission of radiation.

The paternity of the invention of the laser has been the subject of thirty years of patent litigation. On May 16, 1960, Theodore H. Maiman operated the first working laser in Malibu, California, at the laboratories of Hughes Research. It was a solid-state laser that used ruby crystal to produce a red laser beam with a wavelength of 694 nm. Three years earlier, Gordon Gould, who met and discussed with Townes, noted several remarks on the optical use of masers and the use of an open resonator, a detail later common in many lasers. Considering himself the inventor of the laser, Gordon Gould deposited his notes at a notary, but in the legal dispute that arose, he was not recognized by the patent office as the father of the invention [11]. In 1977 Gordon Gould was granted a patent for "optical pumping" and in 1979 a patent described a wide variety of laser applications, including material heating and vaporization, welding, drilling, cutting, distance measurement, communication systems, photocopying systems, and various photochemical applications.

At the end of 1960 the American Iranian physicist A. Javan (1926–2016) made the first gas laser (helium–neon), in continuous emission and in 1962 semiconductor lasers were born. Subsequently, there were conceived laser of various kinds (solid state, gaseous, liquid, semiconductor), which almost continuously cover the entire arc of radiations from ultraviolet to far infrared. Organic dye laser allows to obtain an emission that can be tuned in frequency over a wide spectral range.

The essential elements of a laser are an active medium, a pumping process to make a population inversion, and suitable geometry of optical feedback elements. The active medium consists of a host material (gas, liquid, or solid) containing an active species (see Fig. 2.4). The laser is based on the interaction between atomic systems and electromagnetic radiations.

In LAD the material is ablated by the surface of a target by irradiation with a high-power pulsed laser and collected on appropriate substrate. It is very used for the deposition of semiconductor materials.

Using laser to ablate material must be traced back to 1962 when Breech and Cross [12], used ruby laser to vaporize and excite atoms from a solid surface. Three years later, Smith and Turner [13] used ruby laser to deposit thin films. This marked the very beginning of the development of the pulsed laser deposition technique.

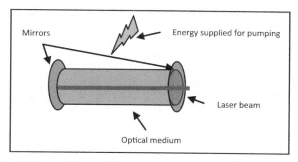

Figure 2.4 Scheme of laser instrument.

PLD is a flash evaporation technique where a small volume of material is heated to a very high temperature and evaporates very quickly so that the composition of the deposit remains that of the source material, named stoichiometry transfer [14].

The interaction of the laser with the target produces a strongly directed emission of material, called plume, which is put on a substrate properly positioned inside the chamber, creating the film. The numerous process control parameters such as wavelength, energy and duration of the laser pulse, beam focusing, target-substrate distance, type of gas in the chamber and its pressure and temperature of the substrate, allow a fine control of the structure, morphology and crystallinity of the film produced, for this reason it is possible to produce both thin films and nanoparticles. The presence of gas, inert or reactive, affects its own dynamic expansion of the plume. Chemical reactions are promoted with reactive atmosphere, with consequent possible formation of compounds even with stoichiometry not known *a priori*. Another important parameter is the pulse duration.

Pulsed laser ablation in liquid, or PLAL, is gaining an increasing interest thanks to several promising advantages, which include environmental sustainability, easy experimental set-up (which does not require extreme conditions of the ambient of synthesis), long-lasting stability of the nanoparticles, which are produced completely free of undesired contaminants or dangerous synthesis reactants [15].

The apparatus is the big advantage, PLAL does not require a vacuum chamber because ablation happens simply in the container of liquid (see Fig. 2.5). Generally vacuum system or pumping gas systems are not need. Laser induced breakdown of submerged targets is characterized by visible plasma emission and the production of shock waves and cavitation bubbles.[12] Upon the bubble collapse, the nanoparticles, which are produced during the plasma cooling phase, can diffuse in the surrounding liquid and form a colloidal solution. We can produce nanoparticles of Au, Ag, alloys, oxides, depending on the target used. The use of surfactant is not required.

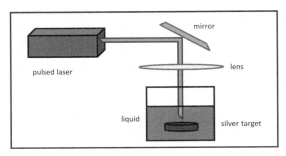

Figure 2.5   Pulsed laser ablation in liquid.

2.2.2   Chemical Methods

In this section we will talk briefly about techniques based on the fabrication of materials from vapor phase and reactions in liquid phase, distinguishing if the reductant is a chemical agent, or a radiation, or an electrochemical method. The choice of the reducing agent depends on the type of nanoparticle that we want to achieve and by operational considerations (green chemistry, reducing cost...).

In any cases, however, the starting point is the reduction of silver ion to metallic silver. The first reminder in the case of solution is the evaluation of E^0 potential standard reduction.

Regarding silver in water, we can see the following reaction:

$$Ag^+ + 1e^- \longrightarrow Ag, \ E^0 = +0.799 \text{ V}$$

Let us remember this value because it will be indispensable in choosing the reductant.

---

[12]Cavitation bubbles: Cavities containing vapors which form when local pressure drops to a value lower than the vapors pressure of the liquid itself, which then passes to the gaseous phase.

2.2.2.1 Chemical vapor deposition

CVD is a technique extensively used in materials-processing technology. It is used to apply solid thin-film coatings to surfaces, but it is also to produce high-purity bulk materials and powders, as well as fabricating composite materials via infiltration techniques.[13] By CVD we can deposit a very wide range of materials.

In chemical vapour deposition the precursors are first vaporized and then introduced in a chamber and adsorbed on a substrate heated at high temperature. The precursor is carried out by a flux of gas like oxygen, argon, hydrogen, nitrogen, and the gaseous decomposition products are also removed from the system [16].

In this technique the parameters that influence the deposition process are the type of precursor, the chemical properties of the gas used as a carrier, the deposition time, the temperature, and the type of substrate.

The precursor is very important, to have a high concentration in the carrier it must be sufficiently volatile, thermally stable, easy to decompose on the surface of the substrate, not too expensive, and should have low toxicity.

We can distinguish two types of precursors: the common inorganic silver salts and the volatile silver (I) complexes. In the first group silver nitrate is the most used in cases such as flame assisted CVD (FACVD), and atmospheric pressure (APCVD) [17].

The volatile silver complexes (I) such as silver(I) carboxylates and their complexes with tertiary phosphines instead are used in classical techniques such as metal organic CVD (MOCVD), aerosol-assistant CVD (AACVD), and plasma enhanced CVD (PECVD) [18].

2.2.2.2 Reducing agents

The choice of the reducing agent is important because it allows to vary the conditions of reaction temperature, the size and shape of the nanoparticles that will be obtained.

The silver value of E^0 allows the use of reducing agents such as hydrazine ($E^0 = -0.230$ V), sodium borohydride ($E^0 = -0.481$ V), sodium citrate ($E^0 = -0.180$ V), hydroquinone ($E^0 = + 0.699$ V), and ascorbic acid ($E^0 = -0.300$ V).

---

[13]Method of composite materials fabrication, a preformed dispersed phase (ceramic particles, fibers, woven) is soaked in a molten matrix metal, which fills the space between the dispersed phase inclusions.

The first reductant we remember and perhaps the most famous is the sodium citrate used in 1951 by the American chemist J. Turkevich (1907–1998) to reduce chloroauric acid $HAuCl_4$ [19].

Later Lee and Meisel used it to prepare AgNPs at temperature close to boiling point of water [20]. Citrate acts both as a reducing agent for the ion and as a stabilizer for the nanoparticles. By increasing the concentration of the citrate, it was seen that the rate of reduction increased [21], the concentration of the reducing agent is therefore important. With this type of reductant, you can obtain AgNPs of about 50 nm in diameter and therefore rather large, the reason is that the reduction is slow. Once the first seeds are formed, the citrate tends to surround them, decreasing the amount of citrate available for reduction. The reduction is demonstrated by the color of the solution that changes from colorless to yellow, green, and gray subsequently.

Another widely used reducing agent is ascorbic acid which we will discuss in more detail in the Chapter 4 (see Fig. 2.6).

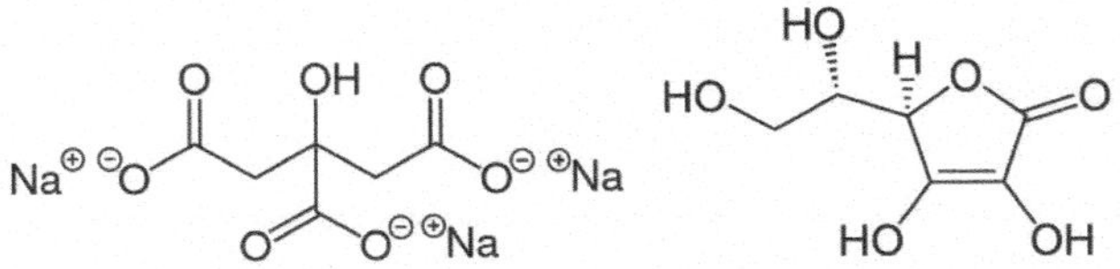

Figure 2.6 (Left) Trisodium citrate and (Right) ascorbic acid.

Moving to a strong reducing agent such as $NaBH_4$, Van Hyning [22] found that it does not follow the mechanism of nucleation and growth of the model of La Mer [23] that is rapid nucleation followed by growth by diffusion. Polti showed [24] that the first step concerns the reduction of ion to silver atoms, after they will build dimers, trimers, and others. Then clusters enlarge to form nanoparticles with a diameter of about 2–3 nm. Particles maintain these dimensions for a certain period, and then regroup into nanoparticles larger than about 8 nm. The stability of the colloid is fundamental at this time. The reducing agent is always in slight excess both to stabilize the formed nanoparticles and because part of the BH_4^- is gradually hydrolyzed. Slowly formed nanoparticles oxidize so that the excess reducing agent also serves to prevent this phenomenon.

Another widely studied reducing agent is hydroquinone (HQ), which acts as a reducing agent by oxidizing to p-benzoquinone (see Fig. 2.7). In this case the concentration of the reducing agent is essential for the size and shape of the nanoparticles [25]. At high concentrations of HQ nucleation is favored while when you decrease the amount of HQ growth prevails over nucleation. This happens when you use nitrate as a precursor, if you use an ammoniacal complex $[Ag (NH3)_2]^+$ growth stops regardless of the concentration of HQ, producing mono-disperse nanoparticles of about 14 nm in diameter. It has been attributed to the fact that the precursor absorbs less on the metal surface of the nanoparticles, so that growth is interrupted.

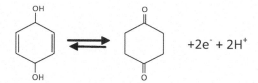

Figure 2.7   Hydroquinone oxidation.

2.2.2.3   Radiation synthesis

The reduction by action of light or other type of radiation is clean and rapid. The photo reduction of $AgClO_4$ by means of UV-light to 254 nm was studied in an alcohol solution by Hada [26]. In this case the mechanism is based on the electronic transfer from the solvent to the excited state of Ag^+ to form Ag (see Eqs. 2.1–2.4).

$$Ag^+ + H_2O \rightarrow Ag^0 + H^+ + HO \qquad (2.1)$$

$$Ag^+ + RCH_2OH \xrightarrow{hv} Ag^0 + H^+ + R\dot{C}HOH \qquad (2.2)$$

$$Ag^+ + R\dot{C}HOH \xrightarrow{hv} Ag^0 + H^+ + RCHO \qquad (2.3)$$

$$nAg^0 \rightarrow AgNP \qquad (2.4)$$

Presence of linear alcohol in solution increases AgNPs formation. Using in this type of reaction a polymer such as polyvinylpyrrolidone (PVP) that works both as a reducing agent and as a stabilizer, the colloidal dispersion is stable for about 6 months [27].

The disadvantage in this type of reaction is represented by the cost of UV lamps and the fact that some molecules used as stabilizers absorb in this spectral region acting as a filter and inhibiting the reduction.

Also, γ-irradiation is known as an effective way for fabrication of AgNPs in aqueous media [28]. Normally the irradiation is done in a solution of $AgNO_3$ mixed with starch, chitosan, poly (vinyl alcohol). The advantage of gamma irradiation method for the synthesis of metallic nanoparticles lies in the fact that desired highly reducing radicals can be generated without formation of any by-product under ambient temperature conditions. Therefore, the radiation technique has proven to be an environmentally benign and low-cost method for the preparation of a large quantity of size and structure controllable metal nanoparticles [29].

2.2.2.4 Electrochemical

The electrochemical reduction of silver is a choice followed by many people. You can start with an aqueous solution of silver salts [30] or using an electrode of silver as the anode and a platinum cathode [31].

In the first case we start with a solution consisting of $AgNO_3$ with KNO_3 as a support electrolyte to the silver salt and PVP as a stabilizer that can coordinate many silver ions (see Fig. 2.8).

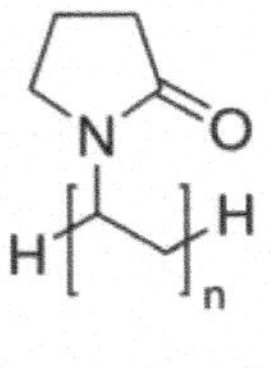

Figure 2.8 PVP molecule, *n* is the number of repetition units.

Two competitive processes coexist in the electrochemical reduction of silver. The first concerns the formation of silver nanoparticles and the second the deposition of thin film of silver on the cathode. In most cases the second process prevails, so it becomes important to reduce it to a minimum as it limits the production of nanoparticles. One solution adopted has been to use PVP as a surfactant and a platinum electrode as a cathode. With this type of method, silver nanoparticles of about 10 nm in diameter are obtained.

The use of a sacrificial silver anode is a much more widely used method, applying an appropriate potential difference with the platinum cathode (working electrode), it is possible in an electrolytic solution the formation of silver nanoparticles. The presence of a surfactant avoids the nanoparticles agglomeration and with consequent deposition on the cathode. The size of the nanoparticles also in this case are about 10 nm.

2.2.3 Biological Methods

We could define biological methods as techniques that involve living organism. The development of nontoxic procedures for synthesizing nanoparticles comprises organisms ranging from bacteria to fungi to plants. Their use is becoming increasingly popular because no reagents potentially toxic to humans and the environment are used. The greatest difficulty lies in finding the right balance between pH, temperature, and other parameters to obtain the desired nanoparticles size and shape.

The reduction of Ag^+ ions by combinations of biomolecules such as enzymes, proteins, amino acids, polysaccharides, and vitamins is ecologically favorable but chemically complex. Many publications report the successful synthesis of AgNPs using bioorganic compounds [32]. Nature provides us with a wide choice of methods to produce AgNPs.

The proteins secreted in the spent mushroom substrate[14] have reduced Ag^+ to provide nanoparticles of Ag protein (core-shell protein) evenly distributed with an average size of 30.5 nm [33].

Do we prefer something spicy? A vegetable, *Capsicum annuum L.*, was used to also synthesize AgNPs [34]. The active substance for silver reduction is extracted by a common juice extractor then it is mixed with an aqueous solution of silver nitrate. The reduction is monitored by UV Visible spectra, the reaction is completed after 11 hours.

If we prefer bacteria, we have a wide choice of possibilities [35, 36].

---

[14]It is the definition of the soil-like material remaining after a crop of mushrooms. This substrate is high in organic matter making it desirable for use as a soil amendment or fertilizer.

A chance may be *Pseudomonas stutzeri* AG259, a gram-negative bacterium. It is an opportunistic pathogen that mainly affects people with impaired immune defenses or physical barriers (skin or mucous membranes). It is the pathogen most often isolated in patients hospitalized for more than a week and one of the microbes involved in the phenomenon of resistance to multiple antibiotics (multidrug resistance). This microorganism can collect metal and producing silver-based single crystals in the size range of a few nm [37].

Another bacterium that we can use in the synthesis of AgNPs is the well-known *Escherichia coli* (*E. coli*), the most known gram-negative bacterium of the genus *Escherichia*. It is an integral part of the normal intestinal flora of humans and other animals. Although most strains of *E. coli* are harmless, there are still some that put human health at risk by causing disorders of varying severity—abdominal cramps, vomiting, and diarrhea with blood. Infection with *E. coli*, which can come from contaminated water or food—especially from foods such as fruits and vegetables, which are often eaten raw, and also from unpasteurized milk and uncooked meat—can be very dangerous especially for young children and the elderly, who can develop a form of life-threatening kidney failure called hemolytic uremic syndrome. *E. coli* is sensitive to heat: cooking food can therefore neutralize it [38].

Talking always about microorganisms but less aggressive than bacteria we find the yeasts that can help us in the reduction of silver. For example, an MKY3 strain extracted from garden soil is able to synthesize AgNPs extracellularly in about 24 hours. The nanoparticles have variable sizes that do not exceed 20 nm [39].

Least but not last, we can find leaves of many plants, *Urtica dioica*, green alga, and *Aloe vera*. Their extracts, aqueous or alcoholic, contain polyphenols capable of reducing silver ion while heterocyclic compounds should act as stabilizers [40].

2.3   Green Chemistry

Chemistry is not always friendly; the indiscriminate use by men of fertilizers, insecticides, herbicides... has made sure that the relationship between chemistry and people is concentrated only on the negative aspects. The so-called green chemistry offers the opportunity to introduce the advantages of chemistry and consequently of new technologies.

Green chemistry is the design of chemical products and processes that reduce or eliminate the use or generation of hazardous substances. Green chemistry is also known as sustainable chemistry, and it applies across the life cycle of a chemical product, including its design, manufacture, use, and ultimate disposal.

In general, green chemistry is a philosophy that spread over to all areas of chemistry, not a single discipline. The aim of green chemistry is to protect and benefit the people, the planet and the economy, by applying this concept in the design, development, and implementation of chemical products and processes.

The way to achieve this goal is to use renewable, and biodegradable materials which do not persist in the environment. The utilization of catalysts and biocatalysts may improve efficiency and conduct reactions at low or ambient temperatures. Chemistry offers a strategic path way to build a sustainable future.

Nowadays, there is a growing need to develop eco-friendly processes, which do not use toxic chemicals in the synthesis protocols. Green synthesis approaches include mixed-valence polyoxometalates, polysaccharides, biological, and irradiation method which have advantages over conventional methods involving chemical agents associated with environmental toxicity. Selection of solvent medium and selection of eco-friendly nontoxic reducing and stabilizing agents are the most important issues which must be considered in green chemistry.

Applying green chemistry means reducing waste generation, avoiding toxic elements and maximizing physical properties, it is the study of materials that are biodegradable, the development of new solvent purification methods, the decrease in quantities looking for alternative syntheses, exploiting reactions that occur at room temperature, thus minimizing energy costs. It is a big challenge (see Fig. 2.9).

We can talk about Green Nanoscience when we apply the principles of green chemistry to nanotechnology. This approach also involves understanding the properties of nanomaterials, including their toxicity for both humans and the environment. The economic advantage in terms of waste disposal and use of raw materials is also important.

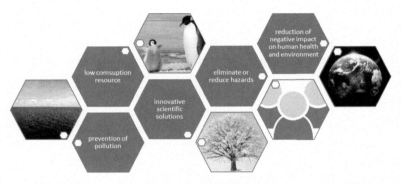

Figure 2.9 Scheme of principal issue of green chemistry.

A growing number of applications sees the use of nanomaterials, for photovoltaics less expensive and more efficient, lighter materials for the automotive sector, nanosensors that have a faster response than common sensors.

Functional groups, nanoparticles can be added to materials already used, reducing energy expenditure and waste generation. Techniques like laser irradiation, and supercritical fluids (i.e., CO_2), are considered green.

When we talk about supercritical fluid (SCF) we mean a low molecular weight fluid with a critical temperature close to the ambient temperature ($T_c \sim 10 \div 40°C$) and a critical pressure not too high ($P_c \sim 40 \div 60$ bar). Light hydrocarbons have these physical properties but have the problem of flammability and toxicity. Certain chlorofluorocarbons are suitable for this purpose but are relatively expensive if obtained with a high degree of purity and are now banned as regards environmental acceptability.

Carbon dioxide, although having a slightly higher critical pressure ($P_c = 72.1$ bar), offers other advantageous properties that make it the fluid most used in SCF applications: non-flammability; environmental acceptability, non-toxicity, low cost even at high purity.

The use of supercritical fluids was initially as a solvent in separation techniques and dates to the last two decades.

The limit represented by the high plant costs has always conditioned its use in the commercial field and only recently the increase in the cost of raw materials, labor, energy resources needed in other processes has made possible their application, together with the advantages of being able to work with reduced volumes and ease of separation of the solvent from the extraction product.

After the first industrial applications of supercritical fluids (decaffeination, purification), the use of SCFs is now employed in the pharmaceutical industry. As regards the production of aromas and essences, a quick analysis allows to assert that, compared to the classic production systems, supercritical fluids extraction (SFE) totally removes the problems associated with the use of organic solvents and the steam distillation. In the first method a portion of solvent remains solubilized in the finished product, in the second one the high temperatures and reactivity of the water are such as to generate changes in the composition of the product, both in quantitative and qualitative form.

The fundamental properties of SFEs can be summarized as follows: solvent power and selectivity of the fluid used; the fluid used is a non-flammable gas having a total environmental acceptability; the operating temperatures make it a "cold" technology; it is possible to modulate the pressure as well as the temperature to optimize both the extraction process and the separation one.

This technique has recently gained attention for the synthesis of inorganic nanoparticles. SCFs such as H_2O and CO_2 are non-flammable, nontoxic, and easily accessed materials [41].

References

1. Lim, B.S., Rahtu, A., and Gordon, R.G. *Nature Materials* (2003) **2**, pp. 749–754.
2. Pallavicini, P., Taglietti, A., Dacarro, G., et al. *Journal of Colloid and Interface Science* (2010) **350**, pp. 110–116.
3. Joule, J.P. "On the mechanical equivalent of heat," *Philosophical Transactions of the Royal Society of London* (1845) **140**, p. 160.
4. Li, L., Chen, W., Zheng, J., Wang, L., and Chen, Y. *Journal of Materials Science and Chemical Engineering* (2016) **4**, pp. 34–39.
5. Boxman, R.L. *IEEE Transactions on Plasma Science* (2001) **29**, pp. 759–761.

6. Kasprowicz, M., Gorczyca, A., and Janas, P. *Current Nanoscience* (2016) **12**, pp. 747–753.

7. Tien, D.C., Chen, L.C., Van Thai, N., and Ashraf, S. *Journal of Nanomaterials* (2010) pp. 1–9.

8. Tseng, K.H., Chen, Y.C., and Shyue, J.J. *Journal of Nanoparticle Research* (2011) **13**, pp. 1865–1872.

9. Einstein, A. *Physikalische Zeitschrift* (1917) **18**, pp. 121–135.

10. Schawlow, A.L., and Townes, C.H. *Physical Review* (1958) **112**, pp. 1940–1949.

11. Gould, G.R. "The LASER, Light Amplification by Stimulated Emission of Radiation" in *The Ann Arbor Conference on Optical Pumping* (Eds P.A. Franken and R.H. Sands), p. 128, The University of Michigan, 15 June through 18 June, 1959.

12. Breech, F., and Cross, L. *Applied Spectroscopy* (1962) **16**, p. 59.

13. Smith, H.M., and Tuner, A.F. *Applied Optics* (1965) **4**, p. 147.

14. Klamt, C., Dittrich, A., Jaquet, B., et al. *Applied Physics A* (2016) **122**, p. 701.

15. Zeng, H., Du, X., Singh, S.C., et al. *Advanced Functional Materials* (2012) **22**, pp.1333–1353.

16. Hitchman, M.L., and Jensen, K.F. (eds.) *Chemical Vapor Deposition: Principles and Applications* (1993), Academic Press, San Diego.

17. (a) Brook, L.A, Evans, P., Foster, H.A., et al. *Journal of Photochemistry and Photobiology A: Chemistry* (2007) **187**, pp. 53–63. (b) Yates, H.M., Brook, L.A, Sheel, D.W. *International Journal of Photoenergy* (2008), pp. 1–8. (c) Spange, S., Pfuch, A., Wiegand, C., et al. *Journal of Materials Science: Materials in Medicine* (2015) **26**, pp. 76–84.

18. (a) Kodas, T.T., Hampden-Smith, M.J. *The Chemistry of Metal CVD* (1994), Wiley. (b) Grodzicki, A., Łakomska, I., Piszczek, P., et al. *Coordination Chemistry Reviews* (2005) **241**, pp. 2232–2258. (c) Gao L., Härter P., Linsmeier C., et al. *Microelectronic Engeneering* (2005) **82**, pp. 296–300.

19. Turkevich, J., Stevenson, J., and Hillier, P.C. *Discussion Faraday Society* (1951), pp. 55–75.

20. Lee, P.C., and Meisel, D. *Journal of Physical Chemistry* (1982) **86**, pp. 3391–3395.

21. Pillai, Z.S., and Kamat, P.V. *Journal of Physical Chemistry B* (2004) **108**, pp. 945–951.

22. Van Hyning, D.L., and Zukoski, C.S., *Langmuir* (1998) **14**, pp. 7034–7040.

23. La Mer, V.K. and Dinegar, R.H. *Journal of American Chemical Society* (1950) **72**, pp. 4847–4854.

24. Polte, J., Tuaev, X., Wuithschick, M., et al. *ACS Nano* (2012) **6**, pp. 5791–5802.

25. Pérez, M.A., Moiraghi, R., Coronado, E.A., and Macagno, V.A. *Crystal Growth Design* (2008) **8**, pp. 1377–1383.

26. Hada, H., Yonezawa, Y., Yoshida, A., and Kurakake, A. *Journal of Physical Chemistry* (1976) **80**, pp. 2728–2731.

27. Guang-Nian, X., Xue-liang, Q., Xiao-lin, Q., and Jian-guo, C. *Colloids Surface A* (2008) **320**, pp. 222–226.

28. Temgire, M.K., Joshi, S.S., *Radiation Physical Chemistry* (2004) **71**, pp. 1039–1044.

29. IAEA-TECDOC-1337,2000. *Proceedings of a Final Research Co-ordination Meeting*, Montreal, Canada, pp. 10–14.

30. Ma, H., Yin, B., Wang, S., et al. *ChemPhyChem* (2004) **5**, pp. 68–75.

31. Rodrigéez-Sánchez, L., Blanco, M.C., and Lopez-Quintela, M., *Journal of Physical Chemistry* (2000) **104**, pp. 9683–9688.

32. Mohanpuria, P., Rana, N.K., and Yadav, S.K. *Journal of Nanoparticles Research* (2008) **10**, pp. 507–517.

33. Vigneshwaran, N., Kathe, A.A., Varadarajan, P.V., Nachane, R.P., and Balasubramanya, R.H. *Langmuir* (2007) **23**, pp. 7113–7117.

34. Li, S., Shen, Y., Xie, A., Yu, X., Qiu, L., Zhang, L., et al. *Green Chemistry* (2007) **9**, pp. 852–858.

35. Bhattacharya, D., and Gupta, R.K., *Critical Reviews in Biotechnology* (2005) **25**, pp. 199–204.

36. Klaus-Joerger, T., Joerger, R., Olsson, E., and Granqvist, C. *Trends in Biotechnology* (2001) **19**, pp. 15–20.

37. Klaus, T., Joerger, R. Olsson, E., and Granqvist C. *PNAS* (1999) **96**, pp. 13611–13614.

38. Gurunathan, S., Kalishwaralal, K., Vaidyanathan, R., Venkataraman, D., Pandian, S.R.K., Muniyandi, J., Hariharan, N., Eom, S.H. *Colloids Surf B* (2009) **74**(1): pp. 328–335.

39. Kowshik, M., Ashtaputre, S., Kharrazi, S., et al. *Nanotechnology* (2003) **14**, pp. 95–100.

40. (a) Jyoti, K., Baunthiyal, M., and Singh, A. *Journal of Radiation Research and Applied Sciences* (2016) **9**, pp. 217–227. (b) Arya, A., Mishra, V., and Chundawat, T. *Chemical Data Collections* (2019) **20,** pp. 100190–100196. (c) *Biotechnology Progress* (2006) **22**, pp. 577–583. (d) Harris, A., and Bali, R. *Journal of Nanoparticles Research* (2008) **10**, pp. 691–695.

41. (a) Meng, Y., Su, F., and Chen, Y. *Scientific Reports* (2016) **6**, pp. 31246–31258. (b) Ji, M., Chen, X., Wai, C., and John L. Fulton *Journal of American Chemical Society* (1999) **121**, pp. 2631–2632.

Chapter 3

Textiles and AgNPs

3.1 Textiles Over the Centuries

The need to cover the body to defend itself from severe winter temperatures or as protection, is a necessity born with the evolution of the human species from the beginning. In the Palaeolithic Age, the time of great glaciations, man hunted animals for their meat, skin, and fur. Thanks to the chipped stone tools with which he cut and scraped skins of animals to make clothes or build shelters.

Archaeological discoveries indicate that the weaving technique, considered the antecedent of the production of woven fabrics, was already known in the Palaeolithic. It was done with the hands, using marsh grasses, grasses, rushes, linden fibers, willow twigs, and others. The simple threads were obtained by twisting of elementary fibers, while the twisted thread consisted in the union of several simple threads subjected to further torsion, a technique with which ropes were used for the construction of huts and palisades, tool handles, belts, or saddlebags. The most ancient type of weaving is linked to fishing activities or to everyday furnishings such as baskets for the transport of foodstuffs and mats.

During the Neolithic Age, the fibers suitable for spinning and weaving were obtained from many plant species, the most used were those obtained from trees such as lime, oak, and elm or those from

Silver Nanoparticles: Synthesis, Properties, and Applications
Anna Facibeni
Copyright © 2023 Jenny Stanford Publishing Pte. Ltd.
ISBN 978-981-4968-21-8 (Hardcover), 978-1-003-27895-5 (eBook)
www.jennystanford.com

herbaceous plants such as flax, hemp, and nettle, and also grasses and rushes.

The discovery of the mummy on the Similaun glacier, on the border between Austria and Italy, found with the remains of the clothes he wore at the time of his death, allowed to deepen the knowledge of researchers on the use of fibers plant [1]. Carbon dating gives it an age between 5300 and 5200 years, placing it in the Copper Age, a time of transition between the Neolithic and the Bronze Age [2]. It is therefore an ancient mummified specimen of *Homo sapiens*. His garments consisted of a knee-length tunic and leather leggings (goat, brown bear, or deer) combined with a loincloth. The clothes had stitching done with animal tendons (especially of ox) and with vegetable fibers. He also had a calf leather belt that contained hunting tools. Numerous ropes mainly made of vegetable fiber mesh and some birch bark containers were found alongside. He also had shoes with an ox leather sole and straw padding, and an oval-shaped cap made of chamois leather. His outfit was completed by a cloak with a weave of vegetable fibers, a technique with which the scabbard of a flint dagger and a hunting net were also made.

We can reconstruct clothing of ancient population thanks to the paintings that decorate the walls of houses, tombs, and figurative monuments. The ancient Egyptians had no need winter clothing, in Egypt the weather is hot and dry, most of the clothes they wore were made of linen that was made from the flax plant. The importance of a person was recognized by the type of dress, then the quality of fabric and the shape of dress depended upon how rich the person was. Both men and women wore a tunic made of white linen, the female dress, however, was longer. Even in Greece the weather was mostly warm then the Ancient Greeks wore light and loose clothing, typically made in the home by the servants and the women of the family. The two most popular materials were wool and linen. Peplos was the typical garment worn by women in ancient Greece. It consisted essentially of a woollen cloth fastened to the side by a belt that forms the typical folds, normally open on one side (the right) and stopped on the shoulder by fibulae, a typical brooch. After the middle of the 6th century BC, the use of the chiton, a dress of oriental origin introduced in Greece by the Ions, was affirmed; of linen or other light material, it was made with a cloth sewn like a bottomless bag, tightened at the waist by a cord and stopped at the shoulders by two buckles. Short for men, long for high-ranking characters and women, it was open on the side or entirely closed. Most of the clothing worn in ancient Rome was made of wool, with sheep bred in Italy. In the case of clothes made from other materials the fabric arrived from Egypt, such as linen, from India such as cotton, and from China such as silk. The tunic was like a long shirt and was the common type of clothing for men. It varied in length from just above the knees to the ankles, a belt was often used to hold it tight. Indeed, the toga was a large piece of cloth around 18 feet long and 6 feet wide and was worn by upper class men outside the home or at official occasions. It was wrapped and draped around the wearer according to the latest style. The cloak completed the clothing in the winter period. Roman women wore a longer tunic, and if they are married also a stole. The *stola* was a long-pleated dress held on by belts, it could be decorated with ribbons and colors. Most Romans wore open sandals made from leather. Other types of shoes included closed boots called *calcei* and open shoes (sort of between a sandal and shoe) called *crepida* [3]. The cloak was fastened on the shoulder by fibulae, a particular type of brooches from the late Bronze Age, found in both male and female burials. Similar objects must have been, according to the Homeric description, the "double-grooved gold clip" decorated with a dog that held a mottled fawn between the front legs, closing the "purple wool cloak" worn by Ulysses or the "twelve gold brooches closed with curved hooks" that adorned the "beautiful peplum" donated to Penelope by Antinoo [4].

So, over the centuries the only ones to be used were natural fibers, this up to the first man-made fibers from the 30s of the 20th century.

Whether it is natural fibers or man-made we are talking about polymers. Jacob Berzelius (1779–1848), a Swedish chemist, recognized the existence of groups of atoms that are repeated within an organic molecule and coined the term polymer in 1833 to describe organic compounds which shared identical empirical formulas, but which differed in overall molecular weight [5]. Given its importance in the chemical field, let us briefly recall Berzelius. His contributions were decisive in all the main fields of chemistry: organic, inorganic, and physical chemistry and also electrochemistry. He determined the atomic weight of the elements obtaining results, with few exceptions, very close to the current ones. He confirmed, with numerous experiments, the validity of the law of multiple proportions of Dalton, fundamental for the atomic interpretation of chemical reactions, extending it also to organic reactions. His contributions in electrochemistry were very important: he was responsible for the introduction of the mercury cathode cell, he compiled a table in which the elements were ordered from the most electropositive (potassium) to the least electropositive (oxygen) and he observed that the more the elements were far in the series the greater their chemical affinity was. In 1803 he demonstrated the possibility of decomposing the compounds by collecting the metals at the negative pole and the "metalloids" (as he called the non-metallic elements) to the positive (the concept of positive and negative electricity had recently been introduced by Benjamin Franklin). He introduced the concept of isomerism by asserting that molecules may differ from each other due to the different arrangement of atoms. He used element symbols as an expression of their atomic weight by applying atomic theory of Dalton. Furthermore, he combined the symbols to represent "simple compounds." Thus, the copper oxide becomes CuO and the zinc sulfide ZnS. Finally, he indicated the number of atoms with an "algebraic exponent" to indicate that in the compound there was more than one atom of that element. These exponents were apexes written in the upper right corner of the chemical symbol; thus, carbon dioxide becomes CO^2 and H^2O water: the modern way of writing chemical formulas was born, the only difference that today the number of atoms is indicated as subscript. He built equipment to develop new purification processes and analytical techniques with the help of Giosuè Vaccano, a glass blower. He discovered a series of elements such as cerium (in 1803), selenium (in 1818), thorium (in 1828), silicon (in 1823), zirconium (in 1824), and tantalum (in 1824) and studied and prepared a large number of their compounds. He discovered racemic tartaric acid (a mixture of the two active forms D- and L-tartaric), ferrocyanides (e.g., ferric ferrocyanide, $Fe_4[Fe(CN)_6]_3$, more known as Prussian blue), and vanadium compounds. He formulated the concept of isomers and studied the phenomena of "contact actions" which he called catalysis. He identified a series of substances that influence the reaction rate but do not enter into the stoichiometry of the reaction, for which Berzelius coined the term catalysts [6].

Polymer noun originates from Greek *polỳs* which means many, and *mèros* which means parts, each fragment is named monomer (*mòno* which means single and *mèros*) it is the smallest part of the fiber.

We can define a fiber like an elongated and filamentous structure, whose origin may be natural or synthetic. A fiber may be spun, and then we can produce a fabric.

We often think of polymers as materials created by man in opposition with the health of the environment. The problem of plastic pollution has increased in recent years when man realized that he had abused a material created not to be disposable but instead of wood, and therefore endowed with a certain duration.

Polymers are the first material with which man came into contact. We can think of meat proteins, of wood used like building material, of natural rubber (Hevea Brasiliensis) used by the people of South America (e.g., Aztecs) to make elastic and waterproof fabrics, to get to wool, silk, cotton used for clothing. A well-known producer is marketing tires made up of mixtures of natural rubbers derived from dandelion root, another time these are polymeric materials.[15]

3.2 Classification

Fibers may be classified among natural, artificial, and synthetic. The textile fibers are the raw material of the textile chain, which includes the processing of the fibers to obtain the yarns, fabrics, and then, finally, products packaged.

The yarn is a collection of textile fibers together and twisted to form a continuous thread, which can then be used to manufacture the fabrics or other textile applications. The fibers can be classified according to their origin. The natural fibers can be of vegetable origin (cotton, flax, hemp....), animal (wool and silk), and mineral (glass fibers and asbestos). Chemical fibers are made from the cellulose of trees (artificial fibers) or oil (synthetic).

We can group the methods for the preparation of silver nanoparticles into two different categories depending on the approach.

---

[15]https://www.continental-tires.com/car/about-us/media-services/newsroom/taraxagum

Fibers may be classified among natural, artificial, and synthetic (Fig. 3.1).

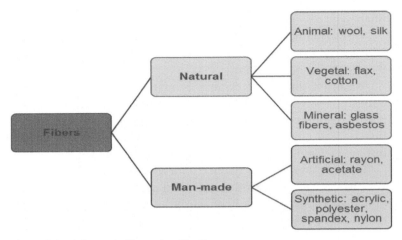

Figure 3.1 Schematic fibers classification.

We will not dwell on minerals because they have never been tested in our research.

We can define the fibers based on:

- **Section** is the transversal or longitudinal conformation typical of each fiber. Natural fibers have a typical defined section, while artificial and synthetic fibers have cross sections depending on the different shape of the hole in the extrusion die and the spinning process.

- **Density** or specific weight, that is, the mass per unit of volume generally expressed in g/cm^3. A low density corresponds to a bulky and light fiber and the corresponding thread or yarn will have a greater covering power.

- **Resumption of humidity** expresses the aptitude of the textile fibers to absorb and retain water. Indicates the hygroscopicity of the fibers: the natural fibers are the most hygroscopic, the synthetic fibers the least hygroscopic.

- **Toughness** is the force compared to the linear mass. Indicates the greater or lesser aptitude of a fiber to withstand traction, also called breaking strength. The toughness can be indicated with different measurement systems: the most common are the g/den, g/dtex, and the cN/tex. Fibers and yarns can be subjected to dry or wet toughness tests. In the case of wet toughness tests, there is usually a accentuated decrease in the toughness of the material, except in the case of vegetable fibers, which show an increase in value.

- **Loss of wet toughness** indicates the difference between dry tenacity and wet toughness expressed as a percentage value.
- **Elongation at break** is the quantified expression of the extensibility of a textile material, that is, of its ability to stretch in the presence of a tensile stress. Elongation at break is the expression in% of the difference between the initial length of the material and its length at the time of breaking.
- **Elastic modulus or modulus of Young** is a measure of the inclination of the initial stretch of the load-elongation curve. It represents the relationship between load and elongation and expresses the force necessary to cause the unitary lengthening of the material. A very high elastic modulus indicates a low deformability of the fiber, which will be rather rigid, resilient, and not very smooth. A low modulus of elasticity indicates a high deformability of the fiber, which will be softer, less resistant, and easily creasable. A fiber with high elastic modulus, high elasticity, and good resilience, allows to obtain textile products characterized by "wash and wear" properties.
- **Elasticity** is the ability of a textile material to recover the initial structure after having undergone a deformation such as a stretch, compression, and flexion.
- **Resilience** is the ability of a textile material to regain its thickness after having been subjected to a certain surface pressure.
- **Crushing or creasing** is the loss of elasticity of a fabric, which tends to no longer recover the initial shape after the deformations suffered, generally resulting from a bending action.
- **Flammability** is the ability of a material to enter and remain in a state of combustion, with the emission of flame, during or after it has been subjected to the action of a heat source.
- **Softening point** represents the temperature at which the fibers begin to soften, becoming sticky.

- **Melting point** represents the temperature at which the polymer passes from the solid to the fluid or liquid state.

3.2.1 Natural

Natural is an adjective which identifies a thing existing in or derived from nature; not made or caused by humankind. So, when we will talk about natural fibers there are vegetable (cotton, linen) or animal (wool, silk), and mechanical processing does not modify their structure.

It seems that the domestication of sheep and goats have started in 7000 BC. In the archaeological site of Tepe Sarab (East Iran) was discovered a reproduction of sheep in clay with a very long fleece (5000 BC) it confirms that sheep with fleeces suitable to give wool for spinning and knitting were already available at that time [7].

The breeding of sheep and wool textile use is often recorded in the Holy Bible, in ancient Assyrian and Babylonian documents, and in the *Odyssey* of Homer (9th century BC). Interesting discovery of woolen fabrics from the Bronze Age have been found in Denmark [8].

In ancient Greece, with reference to the Iron Age, which began around 800 BC, the wool was the most important raw material to produce textiles.

The wool is a polymer of proteic origin and consists of keratin, macromolecule resulting from the union of several amino acids linked together by a peptidic bond. Most of the silk consists of the fibroin fibrous protein and an amorphous gummy protein called sericin, which cements and holds fibroin fibers together. The long chains are composed of a repetition of a segment of six amino acid residues with the sequence: (-Gly(glycine)-Ser(serine)-Gly-Ala(alanine)-Gly-Ala)$_n$. This sequence forms a β structure with the side chains of the Gly that are all arranged on one side of the structure, while the side chains of the Ala and Ser residues are located on the other side of the structure. These β structures assume a microcrystalline disposition in which the faces of the structures containing the side chains of Gly are in contact with each other and on the other side there are instead contacts between the side chains of Ser and Ala.

The fibers of vegetable origin are instead cellulose-based, a polysaccharide with a glucose base (see Fig. 3.2) with a variable content of lignin. Lignin is a high molecular weight macromolecule.

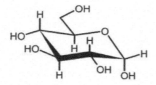

Figure 3.2   Structure of glucose in chair conformation.

3.2.1.1 Wool and silk

The story of wool begins to select sheep for better wool. After carding, Greek women yarned wool flakes, after which they stretched and twisted the fibers. Then wool was spinned with spindle and distaff (remember Moirai and Parcae in Greek and Roman mythology).

Wool was the most common fiber used in ancient Rome yet. The wool was also imported from Attica, Asia Minor, and Gaul. The oldest archaeological finds of textile, in wool fibers, found in Europe and in the Middle East, date back to around 1800 BC [9]. The lack of more ancient finds is due to the bad preservation conditions that led to the complete degradation of tissues eventually produced in previous eras.

The wool is obtained through the shearing operation, cutting the hair, which for the sheep occurs in spring. This wool is called virgin wool. Another method to obtain wool takes place after slaughter. The resulting wool is called tanning wool. The industry also reuses the wool obtained from production waste, in this case we speak of regenerated wool.

The animals from which wool is mainly derived are the merino sheep, a variety selected in Spain around the 12th century. Currently breeds were extensively raised in Australia, South America, and South Africa, it produces a very fine and precious wool. Indigenous breeds have a coarser coat, traditionally used for making mattresses; the goat of angora bred in Turkey, South Africa, United States, from which the mohair wool is obtained, the cashmere goat originally from Kashmir (Tibet) also common in India, China, Iran, Afghanistan from which a very fine wool is made. The alpaca and the vicuna are two types of llamas that live in the Andes. We have also the camel in the Asian continent and the dromedaries in Africa, and finally the angora rabbit.

The processing of wool involved numerous steps involving many craftsmen. The first consisted of washing, after being dried in the shade, the wool returned to the laboratory in which the wool was worked that sent it to other craftsmen assigned to other steps before spinning; one of these was degreasing, obtained by immersing it in special tanks with detergents. From the washing water separates the fat then transformed into lanolin useful for making creams and ointments. The washing phase is still today the one with the greatest impact on the environment, as high quantities of water and thermal energy are consumed; in this phase detergents and chemical agents functional to the operation itself are also used. At this point the wool was ready to be delivered to carders,[16] who frayed the bows in threads of various lengths; the shorter threads would have been used in producing the yarns used as weft and the longer ones for those used as warp during the subsequent weaving process. Spinning was generally assigned to women, who worked at home; the spinning process consisted of two operations carried out simultaneously, the twisting and ironing of the thread, which was wound in a spiral and tensioned using spindles and packages. The spindle was a small stick of iron or wood with an enlarged part in the center, for example, a wheel, on which the fibers were arranged, giving a continuous rotary movement that made it turn on itself, twisting and stretching the threads; the fortress was used to support the fibers to be spun, it was a longer stick generally held between the arm and the side of the spinner. After the pieces had been woven, they were delivered to the wool maker to be washed and dyed.

The wool is water-repellent and has a high elastic recovery, which for practical purposes determines swelling, thermal insulation, wear resistance, recovery of the creases, and above all resilience and durability. Very important from the mechanical point of view is its felting capacity, an exclusive characteristic of wool. This operation occurs when the fiber is subjected to cyclic mechanical forces, in the presence of water, felting is a progressive and irreversible operation. Wool can absorb water vapor up to a third of its weight without giving the feeling of wet. This happens because the amino acids

---

[16]Cardo is a genus of plants belonging to the Asteraceae family, with an appearance of annual or perennial herbaceous plants, of average height, generally very thorny and with similar flowers with artichoke. Their dried inflorescences were used for carding wool to make them softer, smoother, and even cleaner.

that make up the fiber can attract and incorporate water molecules into the structure of the fiber itself, unlike, for example, the sponge, which has a capillary absorption. In the case of humid climate or intense sweating, the wool activates a transpiration process so that it absorbs moisture and returns it to the environment.

Silk is produced by some insects of the Lepidoptera family and by spiders. The silk used to make fabrics is obtained from the cocoon produced by the silkworms, mostly of the genus *Bombyx mori* that secretes to raw silk to protect oneself during growth. *Bombyx mori* is a butterfly species of the Bombycidae family, whose larva is known as a silkworm. The discovery of silk production from the *Bombyx mori* worm occurred around 2700 BC. According to Chinese tradition, the bride of Emperor Huang Ti, named Hsi Ling Shi, invented the first spinning of silk [10]. The breeding of silkworms, known as sericulture, spread throughout China, making silk a consumer commodity highly sought after by other countries. In China, silk was for a long time a privilege of the emperor, the only ones who could wear clothes made with this precious material. Later its use spread among all social classes, and it was also used to produce musical instruments, fish nets, and belts.

Silk Road was the ancient path from China to Europe its name was given by a German geographer Ferdinand von Richthofen [11]. Its length is 4350 miles, from Asia to Europe passing through China, Kazakhstan, Kyrgyzstan, Tajikistan, Uzbekistan, Turkmenistan, Afghanistan, Iran, Iraq, Turkey, Greece, and Italy. Around 6th century AD, some monks brought to Byzantium, the capital of the Eastern Roman Empire, sericulture but only in the 13th century we can find in Europe, especially in Italy, silk production. During the 18th and 19th centuries Europeans progressed in production and already in the 18th century England was at the vanguard in Europe, as a result of innovations in the textile sector. These innovations included better frames and roller printers. In the middle of 19th century, numerous researches on silkworms were carried out that paved the way for a more scientific approach to the production of silk.

The raw silk filament is formed by two fibroin protein pellets, fibrous protein, about 80% by weight, wrapped in a gum called sericin. Under the microscope, the fiber has a regular appearance very similar to that of synthetic fibers. The sericin can be eliminated by treating the silk thread with hot water (degumming): this treatment improves the shine, flexibility, and fabric "hand" of the fiber. The various fabrics made with silk thread are taffeta, georgette, chiffon, organza, satin, velvet, damascus (fabric), brocade, and chinese crepe. Silk is mainly used to produce precious fabrics in the fields of furnishing (curtains and upholstery), clothing (ties, shirts, scarves, and especially women's linens), sacred and liturgical furnishings. It is often mixed with wool, rayon, and other synthetic fibers to improve its quality. Silk fibroin is also studied for possible applications in medicine, to produce prostheses but also for the repair of damaged human body tissues.

3.2.1.2 Flax and cotton

Fertile soil and dry climate of the Egyptian lands allowed to cultivate different kind of plants; among these, flax occupied a special place, because it was used to make clothes worn by everybody. *Linum usitatissimum* is an annual cycle plant with sturdy stems and thin leaves high 30 cm to 1 m with blue or white flower, narrow and pointed leaves. The fiber is obtained from the stems after letting them macerate in water. Nowadays flax is grown mainly in Russia, Holland, France, Belgium, Ireland, and the United States. Flax is formed like cotton almost exclusively from cellulose but also contains fatty lignin waxes. The filaments are 50–60 cm long on average and are formed by elementary fibers 2–4 cm long welded together by sticky substances.

In the site of Çayönü, located 40 km northwest of Diyarbakır, at the foot of the Taurus mountains, was discovered the earliest known of piece of cloth, which was found still wrapped around an antler, it is 9,000 years old and it thought to be a linen fabric, woven from locally grown flax.[17] In ancient Egypt linen was the king of textiles. Since 3000 BC all residents wore linen clothes, the difference was in the styling. Ancient Egyptians used linen also to envelop mummy and preserve them in years [12]. An ancient legend tells that to conceive the linen bandages was Isis, to wrap the body of Osiris, her brother and husband. Egyptians believed that this fiber had sacred and divine origins, it was the oldest fabric, its immaculate color made a symbol of purity, it was the only type of tissue that could be introduced in a temple yet.

---

[17]https://oi.uchicago.edu/research/projects/joint-prehistoric-project

Between 1100 and 1300 AD large flax manufacturing centers arose in northern Europe, particularly in Flanders and Ireland. In Europe, it had its highlight moment between the end of the 1800s and the first decades of the 1900s, with the explorers and great travelers who ventured into Africa and Asia dressed in linen, launching colonial style fashion; it was a tradition for young women in our country, up to half a century ago, especially in the south of Italy, to own linen in the outfit of the linen to bring as a dowry to the wedding. European linen is considered the best and over 70% of the longest obtained fibers, of greater value, are exported, especially to China. There are countries that, however, prefer to opt for short-fiber linen, such as the United Kingdom, whose cultivation is more easily adaptable to the machinery already present in the company. The fiber is soft, flexible, and more resistant than that of cotton (it has a tenacity that varies from 15 to 25 g/dtex against 3–5 g/dtex of cotton), but it has higher production costs, which is why cotton has gradually replaced the use of linen. Like cotton and hemp, wet fiber has a 40% increase in toughness. It is a fiber with low vigor (the linen fabrics crumple easily and do not resume the fold except after ironing), it conducts the heat well: hence the sensation of coolness that the linen fabrics give to the touch (cold hand and slippery). It is an antistatic fiber, like cotton, that is, it does not retain the electrical charges accumulated on its surface. The fiber has a shiny appearance and a recovery of average humidity, and it is insensitive to aging. Currently the main producers of fiber linen in the European Union are France, Germany, Belgium, and the Netherlands. The other big producers in the world, with a production approximately equal to that of Europe, are the Asian countries, and, a little more distant, America. Fiber plants, including flax, are soil improvement, because they have a low need for fertilizers, pesticides, and herbicides, and above all for their root system that develops in depth, making an improvement in the structure and fertility of the land, which the following crops use. Linen fiber is used almost exclusively in textiles. Linen processing waste is used to create panels for thermo-acoustic insulation.

Cotton is another natural fiber, belonging to the Malvacee family, annual herbaceous plant, it is cultivated in tropical countries, for example, United States, Brazil, and Pakistan. Cotton arrives in Europe around AD 1000 unlike countries such as India and Peru where it was already cultivated by the 2nd millennium BC. Arabs took it in Europe, specifically in Sicily, but it was not diffused until AD 1300, because it was very difficult to process.

When Cristoforo Colombo discovered America a new kind of cotton, easier to work, it was found. Over the year's harvests found widespread in French and British colonies of today Southern United States. The appearance of first cotton gin machine in 1792 brought down production costs.

The fruit of the cotton has the shape of a capsule that contains the seeds covered with hair: when it reaches maturity, it opens and lets the cotton bud out. With longer hairs, yarns are made, the shorter ones are used in the manufacture of rayon and celluloid. Oil is extracted from the seeds, livestock feed, and fertilizer. Cotton fiber is made up of almost pure cellulose (95%). The value of a cotton is greater the longer the fibers are (up to 4 cm), also because other properties like shine, fineness, resistance are accompanied by greater length [13].

There are many cotton textile products, but the most famous are the blue jeans, chunky and sturdy cotton trousers originally light blue with showy stitching. The name derives from *blue genes*, meaning blue fabric of Genoa because it was a fabric manufactured in Genoa and exported to the United States where it was used as a work garment.

Hydrophilic cotton is a cotton that has a high absorbency obtained by subjecting the fiber to a chemical project that deprives it of the waxy coating that covers it. Carded and sterilized it is widely used for sanitary and hygienic uses.

Cotton wool carded cotton used for padding. It is formed by numerous layers of cotton, pressed by means of cylinders. Mercerized cotton is obtained by treating the fiber with caustic soda solutions. This treatment is followed by careful washing and drying. The yarn takes on an appearance like artificial silk, increases resistance and absorbs coloring substances more easily.

3.2.1.3   Others

Hemp, *Cannabis sativa*, belongs to the Cannabinaceae family, is characterized by abundant biomass and reaches remarkable productions of dry substance in a relatively short growing cycle. The total dry substance comprises 90% of cellulose and hemicellulose and 4% of lignin. It is not particularly demanding from a climatic point of view but achieves the best performance in warm-humid temperate climates that allow the development of large masses of organic substance. Worldwide, the largest areas planted with hemp, both by fiber and by seed, are found in Asian countries. At the European level, the nation with the largest area of hemp is Russia. Hemp has the paper industry as its main destination (80%), the other uses concern composite materials for the automotive industry (15%), building materials and building insulation (4.5%), and only 1% is for the textile sector (clothing and furniture), including traditional applications (ropes, etc.) and agrotextiles.

Broom, *Spartium junceum L.*, has been known since ancient times for its use as a fiber plant. It was, in fact, already used by Phoenicians, Carthaginians, Greeks, and Romans, to produce mats, ropes, and various artifacts. The same etymology of the Greek word *spartos*, which means rope, confirms the traditional use of fiber for making coarse fabrics. The broom is present in the spontaneous state throughout the Mediterranean basin, from southern France to Asia Minor, is also widespread on the Atlantic coast of Morocco, Portugal, and the Canary Islands. It is a plant typical of the temperate-warm zone, with a mild and humid winter. The resistance of broom yarns is very good when compared to hemp and flax, in fact, the resistance of broom yarns is less than 5% of those of hemp while it is 26% higher than those of flax. It can be used to produce mixed fabrics in various proportions with cotton and other natural fibers. It has always been used for the creation of bags, carpets, bags, belts, hats, curtains, and ropes. Since jute is currently also used in the making of clothes and trousers, it could be assumed a similar use also for broom. In addition to being used in the textile and paper industry, the fibers extracted from the worms can be used in the construction sector for their acoustic and thermal insulation properties that make them suitable also for the realization of insulating panels. The high surface area and the polarity of the broom fibers make them suitable to produce filters. These filters can absorb heavy metal ions, oils, and volatile substances and can be used in the treatment of drinking water and industrial waste as well as in the purification of environments. Finally, the fibers can be used in composite materials with a polymeric matrix. In this application the broom fibers can be an ecological alternative to glass fibers to reinforce some plastic materials.

Nettle is one of the most commonly known and used medicinal species. It is assumed that since the Stone Age some cultivations could even exist, especially for food, both human and animal. As for its use in the textile sector, since ancient times it was used above all the *Urtica dioica* species to make laces, fabrics, and even to make paper. One of the most interesting news is that thousands of uniforms used by Napoleon's army were woven in nettle. However, in Europe, a real production began only in the 20th century [14] when, during the World Wars I and II, the nettle was used to replace the cotton yarn that had become unobtainable. Around 1940 in Germany and Austria about 500 hectares of land were cultivated with nettle dioica for textile use. With the end of the war, interest in nettle disappeared but, starting in the middle of 1990s, the search for fibers with low environmental impact alternatives to cotton, led to a new interest in this species and, above all in Germany, Austria, and Finland, some research projects were carried out on the agronomic aspects of its cultivation and on the technical methods and processes for the extraction of the fiber to be allocated to the textile sector. *Urtica dioica* it is an herbaceous plant, which grows in the temperate regions of Europe, Asia, and North America up to an altitude of 2400 meters. In Europe it adapts to a wide range of climatic conditions and lives in soils with a high content of decomposing organic material (especially rich in N). Nettle fibers have a high cellulose content more than 80%. Fabrics made from nettle fibers are particularly resistant, soft and breathable. The fabrics currently available in Germany are mixed nettle-cotton, with a nettle content varying between 5 and 10%. In the horticultural field, especially in biodynamic agriculture, it is used as a fungicide. In the food field, nettle can be used as all "country herbs"; the most tender parts of the plant are used, that is, the shoots. In the field of cosmetics, it is used to prepare soaps, shampoos, and lotions. It is used as a diuretic, hemostatic, against arrhythmia, anti-rheumatic, and anti-inflammatory. The leaves and softer parts of the plant are preferably used.

Agave, *Agave sisalana,* is a succulent plant of the Agavaceae family, native to the Yucatán peninsula in Mexico, naturalized in southern

Africa, Australia, western India, Madagascar, and Hawaii and in the driest regions of the Mediterranean basin. Often cultivated for ornamental purposes, its textile leaves produce a textile fiber of good quality and very resistant, commercially known as sisal. The name derives from Sisal, Mexican port of Yucatán. Sisal has always been used to produce strings, hawsers, ropes for industrial and agricultural use, and also to produce coarse fabrics, for sacks, carpets, hammocks, brushes, and others. A recent application has extended the use of sisal also to the automotive industry, for car interiors, cat scratchers, lumbar belts, slippers, filters, nets for geotextile applications, and mattresses. In recent years, sisal has been used as a reinforcing agent to replace asbestos and glass wool. The production of sisal does not require the use of pesticides and chemical fertilizers, all waste products from processing can be redistributed to the soil, or even, even if still in an experimental phase, the dried pulp can be used to produce methane. Like the cactus, the agaves survive and produce in a sustainable manner even in less fertile soils and arid regions as they have an excellent resistance to drought and a very limited water requirement.

Jute (also called corcoro) is a natural fiber obtained from plants of the genus Corchorus. As for linen, hemp and kenaf, the textile material for production is obtained from the stem of the plant. Jute is an annual herbaceous plant, from the Tiliaceae family. It is a very versatile fiber, usable for many uses both traditional and to replace synthetic fiber. Jute fiber is suitable to produce very fine and raw yarns, to create fabrics and ropes. The most fragile and short fibers are used to produce a jute fabric called Hessian. The most common uses are for bags, packaging for food products, ropes. The production of paper using jute has given excellent results. It can compete with glass wool as a reinforcing agent in plastics. Technologies have been developed to include jute fiber in polypropylene and applications of bio-composite materials is of increasing interest. Jute fiber coverings for geotextile applications against soil erosion, vegetation consolidation, mulching and paving, even on roads, are increasingly widespread. The largest producers of jute in the world are India, Bangladesh, China, Nepal, and Thailand, which cover 90% of world production. Jute cultivation does not create environmental problems. The production of jute fiber paper has some advantages, such as the lower use of chemical compounds, lower consumption of energy thanks to the lower content of lignin compared to the fiber of the trees, the possibility of using wastewater from treatments for the irrigation. Its products can be easily managed without causing any danger to the environment. Jute is an annual renewable energy resource, due to its high biomass production per hectare.

Kenaf, *Hibiscus cannabinus L.*, is an annual herbaceous plant grown primarily for its fiber content in the stem bark. The genus Hibiscus is widespread and includes about 200 annual and perennial species. it has always represented the most widespread crop to produce fiber from Senegal to Nigeria, although during history it has had other uses, especially in Africa. In Europe the development of this crop is concentrated in the Mediterranean regions with sub-tropical climate for an exploitation especially in the technical fiber production sector. Absorbent material, animal bedding, insulation panels, biocomposites, and energy materials are made with this fiber.

Similarly, the **ramie** represents a fiber used in antiquity, belonging to the Urticaceae family, a native of the Far East. The most favorable climatic conditions for the cultivation of branches are found in sub-tropical climates, not subject to frost in the vegetation period and protected from the winds, with annual rainfall around 1000 mm uniformly distributed during the year. The textile fiber is obtained from the bark; with the inner part it is produced cellulose for extremely valuable paper and finally the fresh terminal portions of the stems and the leaves provide a very nutritious product for zootechnical use.

3.2.1.4   The birth of the loom

The frame is a useful tool for the construction of the fabric obtained through the interweaving of two series of wires, perpendicular to each other, called weft and warp.

The first looms appear in the Neolithic and had a simple rectangular framework built with branches or wooden poles placed in a vertical position on which was placed at the top and perpendicularly to them, a third stick, called *subbio*. From this element the warp threads departed whose tension was obtained through weights, in clay or stone, which are very numerous in the archaeological excavations.

An interesting documentation on its evolution is present in some Greek ceramics of classical age, it is a splendid example of it the Skyphos (a two-handled deep wine-cup) with red figures (5th century BC) depicting Penelope and Telemaco, preserved in the Civic Museum of Chiusi, Italy, in which it is possible to notice a refined frame consisting of two uprights and an architrave. The warp is held in tension with weights and is separated by various crossing bars. We can see the detail of the fabric already built, decorated, and wrapped around a roller (*subbio*). The warp was held in tension by pyramid-shaped terracotta weights with a through hole. The Roman philosopher Seneca (4 BC–AD 65) in his letters to Lucilius [15] reports that at his time the "warp loom" had been replaced by a "two-beam" loom that had spread rapidly and which he considered more refined than the previous one. This type of vertical frame continues to be used for tapestry packaging in the Middle Ages and the Renaissance.

In the 3000–2500 BC the use of horizontal frames on the ground was known, where the tension of the warp threads was obtained thanks to the presence of two beams, one front and one rear. Used for millennia by the Egyptians and the Romans, this type of loom, consists of a frame to stretch the warp threads which are alternately spread by the heddles, from the Latin liciu, lace. The heddles are a series of strings connected to each other which, opening like a ring, welcome the warp thread, allowing it to alternate in order to give passage to the shuttle. In more recent looms the string heddles have been replaced by thin metal sheets with a central hole in which the warp thread is threaded. The liccioli are two horizontal rods that subtend and guide the meshes of the heald. The opening obtained by their movement is called pitch or mouth, it allows the weft thread, wound on a spool and contained in a tapered support, the shuttle, to pass through the warp. At each passage of the weft, which is thrown in one direction and the other, the heddles are appropriately raised and lowered according to a pattern of dividing the warp threads which corresponds to a precise design of fabric. Around the 13th–14th century, although the structure of the horizontal frame remained almost unchanged, the technical knowledge led the man to conceive a mechanism that would allow the weaving of figurative lists along the height of the fabric thanks to a manual program inserted along the warp threads (liccetti).

Between 1495–1496 Leonardo drew and designed a surprising mechanical loom for *tesserae*, totally automatic. The drawing of Leonardo represents the first mechanical frame ever conceived, developed starting from the specimens observed both in Tuscany and in Lombardy. The loom is arranged on two levels: the upper one with the actual weaving devices and the lower one with the organs for unwrapping and wrapping the fabric. A single operator does the job by turning the crank handle, taking care of the good performance of the shuttle, the right tension of the fabric and the possible breakage of the wires. Leonardo represents the brilliant solution to the problem of carrying out the warp and wrapping the finished fabric, maintaining the necessary stringing of the threads and of the fabric, and introducing the flying spool launched automatically by a crossbow system, which would have been used only since 1733. The model (Fig. 3.3) is based on the design found in the Atlantic code [16]. The Museum of Science and Technology of Milan, Italy, has built it and it works. It is one of Leonardo's most beautiful working machines.

In the Renaissance, the construction of the looms became more accurate, allowing the production of complex and refined artifacts embellished by the insertion of silk from China. Thus, the workmanship of fine fabrics such as satin, brocade, damask, and velvet flourish, and the weaving expresses its artistic value also in Europe.

Figure 3.3 Loom of Leonardo.[18]

During the 18th century the loom was gradually mechanized, first with hydraulic energy, then with the use of steam engines, and

[18]Museo Nazionale della Scienza e della Tecnologia "Leonardo da Vinci," San Vittore Street 21, 20123 Milan, Italy. Reprinted with permission.

automated to increase productivity becoming the emblem of the industrial revolution itself. The technological innovations due to numerous inventors are reunited and perfected by the French Joseph-Marie Jaquard (1752–1834), at the beginning of the 19th century. He developed a punched card system for the automatic programming of the movement of the heddles, organizing the execution of very complex drawings with the work of a single weaver (Fig. 3.4).

Figure 3.4   Loom Jaquard.[19]

In the 19th century textile production was further perfected and with the introduction of the automatic spool shift in 1894, the chassis was made completely mechanical.

In the Americas, especially for pre-Columbian populations, the fabric was a very important material. Cotton, wool, alpaca wool (reserved for nobility and for weaving the clothes of the sovereign) dyed or used in natural colors were the fibers most commonly used in the Peruvian area where the art of weaving was taught to both sexes. It was woven with the so-called "belt" loom: a narrow loom, very easy to handle, easily transportable, with the lower stick tied

---

[19]Kindly provided by Telaio Fileria Luigi Bevilacqua, picture Angela Colonna.

to the weaver's waist, with just a strap, and the upper one attached to a tree, a pole or a support, so as to leave the weaver freedom of movement. The disadvantage was that it was not possible to weave larger canvases than the opening of the craftsman's arms. Then there were other types of frames called fixed, among these the vertical one that served that was used to weave works of greater dimensions and complexity such as carpets. In Fig. 3.5, we can see modern looms operating in a small workshop located in the hills of the Marche region (Italy).

Figure 3.5 Operating looms.[20]

3.2.2 Man-Made

Toward the end of the 19th century, we found the first artificial fiber.

The story of man-made fibers began in 1893, with the first patent of Count Hilaire de Chardonnet [17] "Improvements in the manufacture of artificial silk." They are the first steps toward the production of cellulose threads, known as rayon.

---

[20]https://illaboratoriodipatrizia.com/

In the following years it is a succession of inventions that improve the viscose thread and create new cellulosic (or artificial) fibers such as acetate and cupro. The quantities produced, initially modest, undergo a new impulse in the 1930s, when technologies are developed that are suitable for processing, ennobling and dyeing. At the end of the 1930s, with the production of synthetic polymers, the first synthetic fiber was born: nylon (polyamide fiber). In the following decades, other synthetic fibers appeared: polyester, acrylic, polypropylene, elastane, etc., each with its own characteristics and peculiarities. Starting in the 1960s, first almost imperceptibly, then to an increasingly marked extent, artificial fibers give way to the synthetic materials that become dominating the textile market. Currently the artificial and synthetic fibers together represent 60% of the world production of textile fibers, and they are those destined to enjoy the greatest rates of development also for the near future. The first to mention the possibility of manufacturing an artificial textile fiber was the English R. Hooke (1635–1703) who, in Chapter 4 of his book *Micrographia* [18], talks about the possibility of transforming into threads suitable for spinning an artificial material seen by him, resembling silk. In 1734, R.-A. F. de Réaumur (1683–1757) in his "Memoires pour servir à l'histoire des insectes," noting that silk "is not that liquid gum that dries," expressed the idea that it could have been imitated with gums or resins [19]. But the possibility of realizing these ideas arose only after the discoveries of nitrocellulose, also known as guncotton, by a Swiss chemist Christian Friedrich Schönbein (1799–1868) in 1845 and its use in the manufacture of celluloid by American inventor John Wesley Hyatt (1837–1920) in 1865.

In Fig. 3.6, we can see the time evolution of textile fibers.

The artificial and synthetic fibers are man-made textile fibers, derived from compounds existing in nature such as cellulose, oil, water, nitrogen, and other small elements doses. Artificial fibers are obtained from renewable raw materials, such as wood cellulose and cotton linters, and are entirely comparable to natural fibers.

Synthetic fibers originate from different polymers obtained by chemical synthesis and, with their innovative characteristics, represent the evolution of the species. The real advantage of man-made fibers is that they can be programmed according to the specific applications for which they are intended. We can therefore have, depending on the needs, bright or opaque fibers, elastic or rigid, very soft or rough, delicate or ultra-strong, colored or transparent... depending on the field of application.

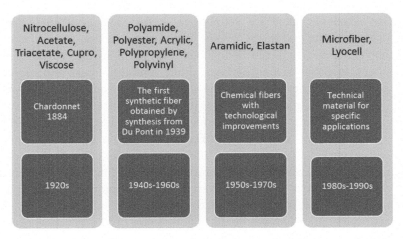

Figure 3.6 Evolution over time of textile fibers.

Titration is the operation that determines the title of a thread or a yarn. Since it is not possible to directly measure the section of a fiber because it is easily deformable and most of the time not circular, the title is used to characterize it.

The value of the standard determines the fineness and consequently the quality of the fiber. The title is a relationship between length and weight; the relationship between weight and length is called direct titration, while the relationship between length and weight is called indirect or numbering titration.

Measurements units of direct titration are:

- Tex = weight in grams of 1,000 m of wire (used in continuous burr fibers and for the flock of chemical fibers)
- Decitex (dtex) = weight in grams of 10,000 m of wire (Tex multiple)
- Denier (Td or den) = weight in grams of 9,000 m of wire (same as Tex)
- Scottish (Ts) = weight in pounds of 14,400 yards of thread (used for jute yarns)

Measurements units of indirect titration or number are:

- Metric (Nm) = meters of yarn in 1 g (used for combed wool yarns, carded wool, chemical fiber staple, fancy yarns)
- Kilogrammetric (Nk) = meters of yarn in 1 kg (used for yarns of waste, carded wool, fancy yarns)
- English cotton (Ne) = skein (840 yards) of thread in 1 pound English (454 g)
- English linen (Nl) = skein (300 yards) of thread in 1 pound English
- English worsted wool (Nw) = hank (560 yards) of thread in 1 pound English
- English carded wool (Ns) = skein (256 yards) of thread in 1 pound English

The most popular measurement of thickness of a female stocking is the denier, abbreviated as den. Then we can define a 10 den as a wire of length 9000 meters that weighs ten grams. It's a very elegant stocking.

3.2.2.1 Artificial fibers: rayon

Rayon is an artificial textile fiber that is obtained by transforming, with chemical and mechanical processes, cellulose into filaments that can be used industrially. For some external features, rayon resembles silk, from which, however, it differs in its chemical composition, resistance, and other physical characteristics. This similarity suggested calling it artificial silk during the start-up of the industry, a name that began to be replaced by the current one in 1924 to avoid confusion and to highlight that this new textile fiber is not just a surrogate but possesses property and special merits. The term rayon, ray in French, derives from the allusion to the extreme brightness of the fiber, comparable to that of the solar rays.

The merit of having brought the production of artificial silk from the experimental phase to the industrial phase is up to the count L.-M.-H. Bernigaud de Chardonnet who at the École Polytechnique had been a pupil of the Pasteur when the latter performed his famous researches on the diseases of the silkworm. The Chardonnet made the first attempts at manufacturing in 1878. After 6 years of research, on 12 May 1884, he presented a memoir at the Académie des sciences, *Sur une matière textile artificielle ressemblant à la soie,* which summarized all the essential elements of his methods from the chemical and industrial point of view. On November 17th of the same year he took the first patent. The process had a very serious defect: the filaments were flammable and explosive. Chardonnet, to reduce its flammability it submitted the yarn to a denitrification process transforming the nitrocellulose back into cellulose. Chardonnet established in 1884 in Besançon (his native country) a company with 6 million of capital to produce his artificial silk, he worked tenaciously perfecting the process and creating spinning machines that were the prototype of those currently in use. At the Paris exhibition of 1889, he exhibited his first machine and the first samples of the new textile fiber. The invention aroused much curiosity and was the subject of a favorable report by the jury.

Currently there are four main types of rayon: rayon with viscose, acetate, cuprammonium, nitrocellulose (also called rayon Chardonnet), which differ both in the manufacturing process and in the characteristics of the finished product. The raw material for the manufacture of viscose rayon is usually wood cellulose; for rayon with acetate and for those with cuprammonium and nitrocellulose, the linters of cotton are used, which are made of almost pure cellulose. The manufacturing process in its general lines consists in transforming the cellulose of these raw materials into substances that give a dense and viscous solution from which the filaments are formed by passing them through very thin holes and immediately after making it coagulate by evaporating the solvent or action of specific solutions.

The tenacity of rayon threads is measured by the ratio between the load under which the break occurs and the area of the section; therefore, it is expressed in grams for money. Another important feature is the lengthening that the wire can undergo without breaking and that is the ratio between the length that the wire has at the time of breaking and that which it had when it was not under tension. This elongation is much higher than the elasticity limit, that is, to that lengthening below which the wire returns to its original length as soon as the tension ceases. Thus, that is, a rayon with acetate may have a 30% elongation and a 3% elasticity limit only. In order for the thread not to break too easily in the weaving process, the elongation must not be less than 10%; but to avoid defects in the fabric, it is good that it does not exceed 20%.

Rayon filaments have a very irregular cross section because their outer layers, which coagulate and harden first, shrink as the inner layers in turn pass into a solid state, contracting. The filaments of each factory have a characteristic section, which one tries to keep constant also because each variation of it translates into a different behavior of the yarn to the dye. To increase their covering power and lower their specific weight, rayon filaments are made that contain gas bubbles, possibly combined to form an internal channel. These bubbles are obtained by mixing the solution to be spun with substances (e.g., sodium bicarbonate) that develop gases that do not attack the rayon or pass to the vapor state when the filament is heated (like some petroleum derivatives).

3.2.2.2 Synthetic fibers: nylon, polyester, acrilic

Nylon is a family of synthetic polymers, precisely the polyamides (amidic structure, Fig. 3.7), whose progenitor nylon 6,6 was synthesized for the first time on 28 February 1935 by Wallace Carothers at Du Pont in Wilmington, Delaware (USA) [20]. There are several types of nylon, depending on the starting reagents, nylon 6,6 is the product of the condensation polymerization of hexamethylenediamine and adipic acid, while nylon 6 is obtained from ε-caprolactam. Nylon 11 obtained from ricinoleic acid contained in castor oil is also industrially interesting.

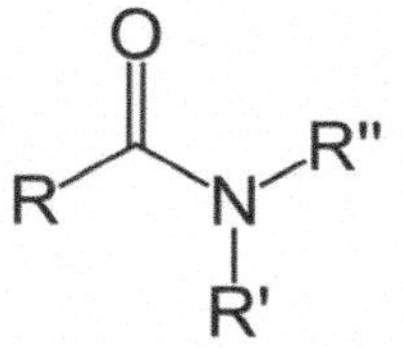

Figure 3.7 Amidic structure (R, R′, R″ = H, aromatic group, alkylic...).

With rayon, modern hosiery began during the World War II, and it was also used in the production of parachutes.

The figure that accompanies the word nylon refers to the number of carbon atoms existing in the single or double component of the elemental molecule. In Fig. 3.8(a), we can see six carbon atoms, instead in (b) we can observe two different carbon unit separated by red dotted line.

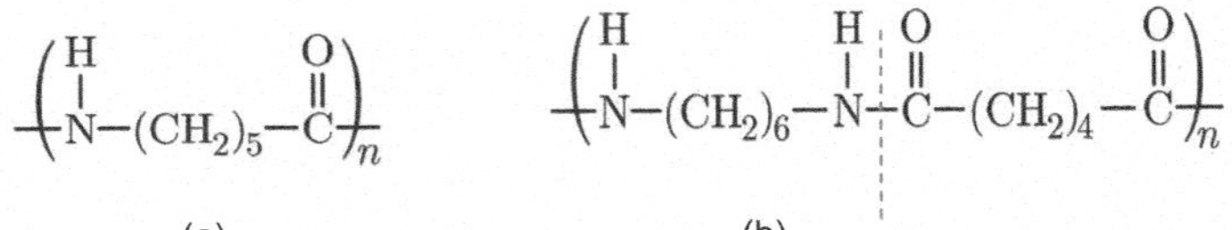

$$\left(\!\!\begin{array}{c}H\\|\\N\!-\!(CH_2)_5\!-\!\overset{O}{\overset{\|}{C}}\end{array}\!\!\right)_{\!n}\qquad\left(\!\!\begin{array}{c}H\\|\\N\!-\!(CH_2)_6\!-\!\overset{H}{\overset{|}{N}}\!-\!\overset{O}{\overset{\|}{C}}\!-\!(CH_2)_4\!-\!\overset{O}{\overset{\|}{C}}\end{array}\!\!\right)_{\!n}$$

(a)   (b)

Figure 3.8 Nylon structures (a) Nylon 6 and (b) Nylon 6,6.

The main features of this fiber are excellent resistance to wear, high elastic recovery, dyeing facility, good color fastness, and ease of maintenance.

The spinning operation is carried out at a temperature of 30–40°C, higher than the melting point of the polymer; to avoid phenomena of depolymerization and degradation it is important that the mass has a moisture content not exceeding 0.1%.

It is sensitive to various chemical reagents (bleaches and mineral acids); not very resistant to high temperatures (>100°C) and to specific environmental conditions such as light and atmospheres rich in nitrogen oxides. In polyamides the numerous intra- and intermolecular hydrogen bonds due to the presence of –CONH groups give rise to intense cohesive forces which, together with the regularity of the chains, lead to considerable percentages of crystallinity. This gives the material excellent mechanical characteristics: high elastic modulus, hardness, and abrasion resistance. The melting point is generally high: 220°C for Ny-6, 262/264°C for 6.6, and 174°C for Ny-12; the glass transition is observed around 50°C for Ny-6. Polar amidic groups, in addition to making polyamides rather hygroscopic, also greatly improve the impact resistance of the material. This is because the adsorbed water molecules act as a plasticizer, increasing toughness.

The nylon is worked to complete the alignment of the molecules and to give the yarn its special characteristics from an aesthetic, tactile, and functional point of view: in particular the most important processes are ironing and texturizing, the wires thus obtained are sold as finished product or can be used, in the case of texturized, for processes in which they are coupled with elastomer.

The yarns are distinguished according to the processing to which they were subjected and to their size; the thickness of the yarn, due to its irregular and compressible nature, is not expressed as a diameter but as a mass per unit length of a suitably tensioned yarn.

The nylon 6,6 threads are generally composed of several burrs (bave[21]) so the textile threads are indicated with two numbers "x/y" of which the first represents the titer of the yarn expressed in dtex and the second the number of burrs that compose it, which correspond to the number of holes in the supply chain from which they come; if necessary a third digit is added to indicate the number of "barre," that is of bunches, for example, a 22/20 ×2 will be a thread composed of two bundles of twenty burrs each for a total of 44 dtex.

The x/y ratio indicates the size of the burrs from which the softness of the fabric follows. For example, a 17/2 has a ratio of 8.5 and a fabric made with this yarn is hard and crepe, as the ratio decreases, we have the "multi-bore" such as 78/68 and finally, when x/y ≈ 1, the "microfilament."

Since burrs can also be very numerous, it is advisable in some way to keep them compact through a binding, then proceed with the "air binding" which consists in blowing air from the nozzles in order to make the burrs casually tangled; areas called "interlacing points" are formed which, based on their quantity, keep the wire more or less united.

There are also aromatic polyamides, known as Kevlar and Nomex (Fig. 3.9).

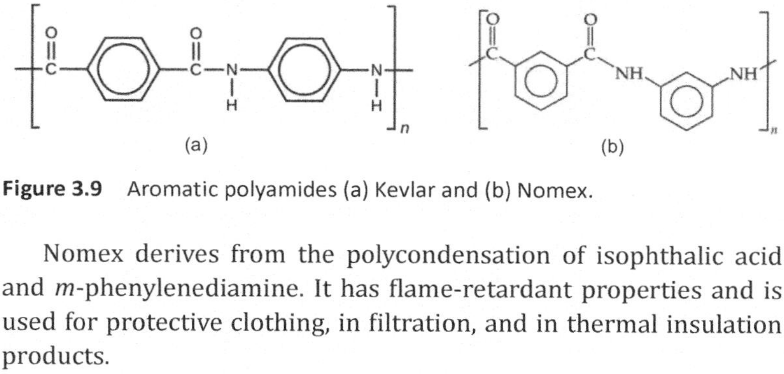

Figure 3.9 Aromatic polyamides (a) Kevlar and (b) Nomex.

Nomex derives from the polycondensation of isophthalic acid and *m*-phenylenediamine. It has flame-retardant properties and is used for protective clothing, in filtration, and in thermal insulation products.

Kevlar derives from the polycondensation of terephthalic acid and p-phenylenediamine. Its main characteristic is its resistance to traction and impact. It is used in aircraft composites, for sports equipment, or in the building sector.

---

[21]Continuous thread

The most common among polyesters, synthetic polymers derived from the polycondensation of a carboxylic acid and an alcohol is polyethylene terephthalate known as PET (see Fig. 3.10). Its properties make it possible to give products in polyester quality such as crease resistance, resistance to wear, dimensional stability. Even after many washes, polyester fabrics do not require ironing and can easily withstand dyeing processes. They are waterproof and resistant to dirt, thanks to their low liquid absorption coefficient. Polyester is often used in combination with other fibers and, to give its surface additional mechanical characteristics, it can be subjected to specific treatments, for this reason it finds application in many sectors. In addition to high stability, if immersed in water, it loses some points of toughness when it is wet but recovers its original characteristics after drying, while in boiling water, especially if in the presence of high-pressure steam, the tendency is more pronounced. It has a good resistance to dry heat up to 180°C, with a melting point above 250°C. If subjected to flame, it tends to melt before burning. The alkali resistance is generally good, but at room temperature the polyester is subject to a slow hydrolysis, which is accelerated to higher temperatures. Polyester is resistant to the most common hydrocarbon solvents, aromatics and chlorinated. The woven polyester combines the characteristics of strength and elongation of the nylon with the characteristics of the form of the rayon.

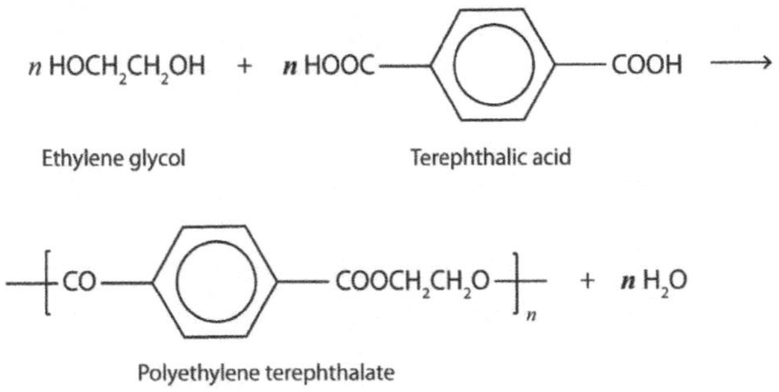

n HOCH$_2$CH$_2$OH + n HOOC—⬡—COOH ⟶

Ethylene glycol   Terephthalic acid

—[CO—⬡—COOCH$_2$CH$_2$O—]$_n$ + n H$_2$O

Polyethylene terephthalate

Figure 3.10 Condensation reaction of PET.

Polyester fabrics can undergo treatments with polyurethane, with Oleo Hydro Repellents and RFL (resorcinol-formaldehyde-latex) substances for impregnation or spreading (single and double-face) to facilitate adhesion to implement their characteristics.

Polyester fabrics are often used in the clothing, furnishing sectors (curtains, flooring, coverings, etc.), and above all to produce technical textiles (transport, geotextiles, medical, safety devices).

The term acrylic is used in the textile industry to indicate synthetic fibers produced from acrylonitrile, $CH_2CHC\equiv N$ monomer which constitutes at least 85% of the repeating units in the polymer chain. The first acrylic fiber to be produced was the Orlando® in 1950 by Dupont.[22] Subsequently, the fiber was marketed under the Dralon® brands by the company Bayer (Germany) and Leacril® by Montefibre (Italy). The monomer is dissolved in an organic solvent (dimethylformamide) and then the solution thus obtained is spun dry. The yarns obtained from the bow, fiocco[23] have the appearance of wool, while the fiber produced in continuous filament resembles silk. When the acrylonitrile content is between 35% and 85%, according to the ISO (International Organization for Standardization) definition, the fiber takes the name of modacrylic.

The main characteristic of the modacrylic fiber is that it is intrinsically flame retardant, and this property is obtained thanks to the introduction into the polymeric chain of specific co-monomers with flame-retardant function: the most suitable are the halogenated products, with which acrylonitrile polymerizes easily. Finally, the acrylic fiber is defined as "homopolymer" when it is made of 100% acrylonitrile: in this case, due to the absence of the co-monomer mentioned above, the fiber is not dyeable, but has mechanical and resistance properties to the very high hydrolysis, which makes it suitable for specific technical uses. Once polymerized, acrylonitrile is spun, dry, as in the first production of DuPont and Bayer, or wet, a system that over time has become the most widespread, representing today about 85% of the installed capacity in the world. Spinning is followed by a series of treatments whose purpose is to impart to the fiber the necessary physical-mechanical characteristics, such as toughness, modulus, fixing (in the case of stabilized fiber), retractability (in the case of high-bulk fiber), hand textile: properties that are designed according to the processing technology and end uses to which the fiber is destined.

The dyeing can be carried out directly in the production phase, following two process methods: spinning, inserting dyes or pigments

---

[22]http://www.dupont.com/corporate-functions/our-company/dupont-history.html
[23]State in which the fibers are found before spinning in bulk, not oriented in parallel.

in the polymer solution, before feeding it to the dies, or in line, adding special dyes before drying, when the fiber it is still in the gel state, so color penetration is effectively possible.

The "producer dyed" acrylic fiber has established itself for a series of considerable advantages, such as the excellent reproducibility of colors between the different dyeing baths, the very high solidity and durability of the colors, the absolute consistency of characteristics, the advantages of cost in relation to dyeing in yarn and finally respect for the environment, because the dyes are not discharged with dyeing water but recovered and disposed of as part of the production cycle.

3.3 Improve the Performance

Smart material is a generic term for a material that in some way reacts to its environment. Smart materials can be classified in many various ways, for example depending on their transforming function: property change capability, energy change capability, discrete size/ location or reversibility. Smart materials can also be classified depending on their behavior and function as passive smart, active smart, or very smart. Another way of classifying them is to look at the role they could have in a smart structure, as sensors or actuators. According to the manner of reaction, they can be divided into passive smart, active smart, and very smart materials. Passive smart materials can only sense the environmental conditions or stimuli, for example, sensors. Active smart materials will sense and react to the condition or stimuli; besides the sensor function, they also have actuation characteristics. Very smart materials can sense, react and adapt themselves accordingly. An even higher level of intelligence can be achieved from those intelligent materials and structures capable of responding or activated to perform a function in a manual or pre-programmed manner. E-textiles, also known as electronic textiles or smart textiles, are fabrics that enable digital components (including small computers), and electronics to be embedded in them.

Textile materials are in common use, both in the field of clothing and in furnishing. Over the years, even the most traditional textiles are required to have performance with increasingly higher technical content: easy-care fabrics, antibacterial and fresh fabrics, breathable fabrics, etc. accompany a tactile aesthetic impact that is valid for the market, with an ever-increasing functionality; the areas of overlap between technical textiles and fashion textiles are always greater.

Think of building fabrics used to reinforce concrete, or bullet-proof clothing for the military, many components for cars and airplanes are reinforced with fabrics, to conclude with applications in the medical field. Textile materials offer many advantages, first of all the flexibility can be draped in different shapes depending on the materials they are breathable, and they are generally light. The turning point came with the advent of synthetic fibers, when scientists were able to produce tailor-made fibers for a certain application.

Think of rescue systems, slides from aircraft, accessories (filtration, transmission belts, battery separators, ropes, etc.) are generally textile reinforcement composite materials, which replace, in whole or in part, traditional materials, such as metals, alloys, plastic materials.

The choice of fibers is dictated, as in textiles in general, by aesthetic requirements, workability, performance, and cost. Man-made fibers and polyester especially dominate the furnishings, but polypropylene is also entering significantly. For tires, for example, after the domination of the more traditional fibers in the past, high-tech fibers, such as aramid fibers, are used in small applications, which increase safety by reducing the weight of the tire significantly. For the protection of man, we range from suits for dangerous jobs (firemen, welders, soldiers...) to work gloves, bullet-proof vests, hospital clothing (barrier fabrics in the operating room...) to protection against electromagnetic waves.

Sport and free time are becoming increasingly important and define lifestyles and therefore also clothing and the use of textiles. Much of sportswear must have significant technical performances, for example, anti-friction fabrics for competitive sports, from bike speed to swimwear, for example, protective and breathable fabrics for skiing and mountaineering. But also, many objects used in sports have a strong textile content, for example, rackets and bats in textile composite, surfboards, and sails. Even camping tents are typically produced in textiles.

A textile treatment must be stable for a few months, with storage at ambient temperature. Aqueous systems are preferable to using flammable solvents, a continuous process is better for companies and the concentration of material to be applied must not be too high.

Many fibers used in textiles have insulating properties, the values expressed in ohm[24] vary from 10^{14} for a polyester up to 10^8 for cotton. The low resistance of the cotton is due to the hydrophilicity of the fiber, the moisture absorbed by the material makes it a good conductor.

The term finishing refers to the set of processing operations that are applied to the fabrics to improve its appearance, hand textile, properties, and possible applications. Finishing operations can be carried out, either through mechanical action, or with the use of chemical substances, or with the use of resins or silicones in the form of microfilm, all with the aim of bringing quality and characteristics to the various textile materials such as to guarantee an optimal behavior in packaging and during use and which play a role of great and growing importance, for the commercial success of the product.

Mechanical treatments are, for example, calendering, that is the passage of the fabric on two or more heated cylinders which rotate at the same speed or at different speeds, which serves to achieve different effects. Normally it is carried out to soften the fabric, or to have a permanent glossy or semi-glossy effect. The sanforization instead makes the cotton fabrics unshrinkable, the fabric is washed, pressed, and dried. An example of chemical treatment is the antibacterial one, the fiber is first dipped in an antibacterial solution, and dried to evaporate the solvent.

The finishing may be applied by simple dipping, by padding, and by spraying.

In dipping the textile is immersed in the liquid for a certain time, then it is extracted and dried in air at room temperature, or thermally.

In padding or foulard technique, all the phases take place continuously, the fabric is initially immersed in a bath and then squeezed into a scarf to remove excess liquid.

In the spraying process, solvent evaporation does not occur after deposition but during spraying itself.

---

[24]Unit of electrical resistance in the SI has the value of 1 ohm when a potential difference equal to 1 V generates a current of intensity equal to one ampere.

Plasma or corona treatments are rather recent methods, useful to sensitize the fabric for specific treatments, specifically to improve wettability.

The functional groups present on the fiber influence the type of possible treatments to be applied (Table 3.1).

Table 3.1   Functional group available

Fiber	Functional group	Behavior
Cotton and cellulosic	$-OH$	Hydrophilic
Wool/Silk	$-OH, -NH_2$	Hydrophilic
Polyamides	$-OH, -NH_2$	Hydrophilic
Polyesters	$-COOH, -NH_2$	Hydrophobic
Polyacrilonitriles	$-CN$	Hydrophilic

Below some treatments are discussed in detail.

3.3.1   Increased Fire Resistance or Flame Retardance

Combustion is a chemical reaction that develops heat, often accompanied by flames and smoke. To obtain this reaction three elements are necessary: the fuel, the oxidizing agent (usually oxygen), and the heat, which together form the so-called fire triangle. A fireproof material is a material that cannot react to fire or do it with extreme slowness, thus preventing the birth of a fire or greatly limiting its expansion. The name fireproof is the translation of Italian word *ignifugo* that derives from the Latin: *ignis* stands for fire and *fugio* is about to escape, but also, to put to flight, to repel.

Fire behavior is evaluated from multiple points of view and then summarizes an overall structured framework (Fig. 3.11).

Fire-retardant fabrics can belong to two macro-groups: they do not burn when exposed to the flame or, more frequently, they burn only when the temperature exceeds a certain threshold and or they burn slowly, hindering the burning of the fire. For this reason, we often talk about fire-retardant fabrics.

Fire-retardant fabrics can have two roots: they are fabrics that are in themselves flame retardant, namely fireproof due to their composition and chemical structure, or they are typically inflammable fabrics subjected to special fireproofing treatments which consist in the addition of substances such as ammonium sulfate, ammonium salts, sodium borate, etc. Some of these substances, when exposed to flame, vitrify the fabric or have it charred without generating flame or generate carbon dioxide or ammonia, so they not only do not propagate the fire but can even reduce the flow. A naturally fire-retardant fabric has the advantage of not having undergone treatments, thus resulting in no applied surface substances that possess toxic potentials. The flame-retardant properties do not therefore lapse either after maintenance or washing, or because of the conditions of use.

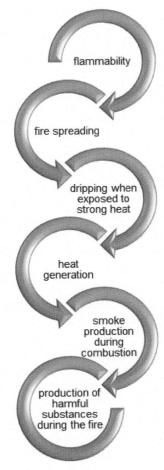

Figure 3.11 Fire behavior of a material.

Instead in the case of highly flammable products, made fireproof through treatments with flame-retardant products such as aluminum, magnesium, boron, red phosphorus, nitrogen (called inorganic flame retardants), with the role of interrupting the self-sustained combustion cycle, when the fire occurs propagates strongly, combustion cannot be interrupted but only confined. The fireproof treatment loses effectiveness after several washes and after a certain time period.

Another problem of flammable fabrics transformed into fireproof, is that their burning can generate toxic fumes, which disturb the sight and breathing; on the contrary fireproof fabrics do not generate any toxic smoke. This aspect is very important because during fires often there are victims not because of the flames but the fumes. If cellulose-based fabrics that burn very quickly, are mixed with synthetic fibers, the result is that the fabric can stick to the skin.

Brominated flame retardants (BFRs) are mixtures of chemicals that are added to a wide range of products, even for industrial use, to make them less flammable. They are often used in plastics, textiles, and electrical/electronic equipment. The most common are polybrominated biphenyl ethers (PBDE), used for plastics, textiles, electronic bodies, circuit systems; hexabromocyclododecanes (HBCDD), in thermal insulation in buildings; tetrabromobisphenol A (TBBPA) and other phenols for printed circuits, thermoplastics (especially in televisions); polybrominated biphenyls (PBBs) for consumer devices, textile articles, plastic foams.

These BFR classes have been marketed as technical mixtures, with different trademarks, consisting of various chemical compounds for each class. In the European Union (EU) the use of some BFRs is prohibited or restricted;[25] however, due to their persistence in the environment, these chemicals continue to raise concerns about the risks they pose to public health. Products treated with BFR, still in use or in the form of waste, leave BFRs "filtered" in the environment and contaminate the air, soil, and water.

Organophosphorus products, especially phosphate esters, represent another category of end-of-life retardants, and represent about 20% by volume of FR production. Although characterized by an efficiency generally lower than that of halogenated systems, numerous phosphorus-containing compounds are used as flame

---

[25]https://www.efsa.europa.eu/en/topics/topic/brominated-flame-retardants

retardants and have gained interest in the last decade due to the growing demand for halogen-free systems. The organic molecules used have very variable structures (e.g., alkyl and aryl phosphates, phosphonates, phosphinoxides, etc.) and complex, while among the inorganic compounds the most used are red phosphorus and ammonium polyphosphate.

There are also flame retardants based on nitrogen, more precisely melamine based (see Fig. 3.12). The action of delay to the flame of the melamine is due to the endothermic decomposition which it undergoes when it is subjected to heating (250 ÷ 400°C). From decomposition there is the development of ammonia (NH_3), hydrocyanic acid (HCN), nitrogen oxides (NO_x), and others irritant products.

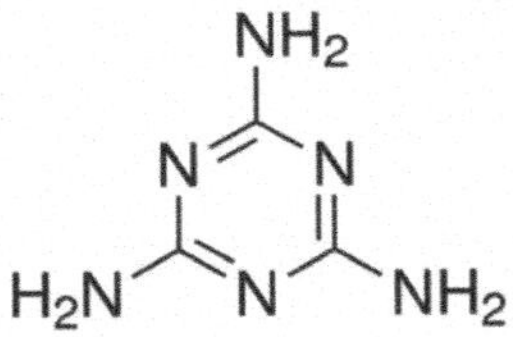

Figure 3.12 Melamine structure.

3.3.2 UV Protection

Sunlight is composed of electromagnetic radiation with many different wavelengths, the visible spectrum ranges from about 400 to about 70 nm, the most damaging range is between 200 and 400 nm (see Fig. 3.13).

The extraterrestrial solar spectrum contains UVC radiations (from 100 to 280 nm), UVB (from 290 to 315 nm), and UVA (from 315 to 380 nm).

At an altitude of about 15–30 km above its Earth's surface there is a gas that serves to shield the harmful radiation UVC emitted by the sun. This gas has the name ozone, a highly reactive molecule that contains three oxygen atoms, is formed and constantly decomposes in the upper atmosphere at 10–50 km of altitude, in the region called the stratosphere.

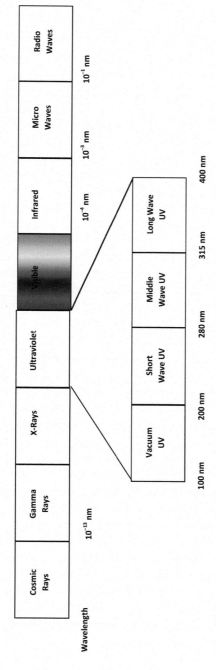

Figure 3.13 Electromagnetic spectrum.

Chlorofluorocarbons (CFCs), chemical products often found in spray aerosols widely used by industrialized nations in the last 50 years—are the main culprits in breaking the ozone layer. When CFCs reach the upper atmosphere, they are exposed to ultraviolet rays which cause them to break down substances that include chlorine. Chlorine reacts with oxygen atoms and breaks down the ozone molecule.

In 1987, to deal with the destruction of the ozone layer, the international community defined the Montreal Protocol on substances that deplete the ozone layer. The goal of the protocol is to reduce the production and consumption of ozone-depleting substances in order to reduce the level of it in the atmosphere and thus protect the ozone layer around the Earth. The global consumption of substances that deplete the ozone layer has been reduced by about 98% since countries began to take measures under the Montreal Protocol. As a result, the atmospheric concentration of the most aggressive types of these substances is decreasing and the ozone layer is showing the first signs of recovery.[26]

When exposed to UV rays, the skin produces a shield consisting of a dark substance called melanin. The result of this defense is known to all because it comes in the form of a "tan." All the radiations emitted by the sun (infrared, visible light, ultraviolet) are fundamental for our health. The body needs ultraviolet to synthesize vitamin D, necessary for the bone structure. A high dose of UV rays, however, increases the risks to the skin: and therefore, the skin tries to defend itself with melanin. For years now, the risks of contracting serious diseases (melanoma) due to ultraviolet solar radiation have been well-known.

Some adults and many children have a photosensitivity to ultraviolet rays and must therefore be careful when exposed to the sun. There are also professions that involve exposure to UV rays, such as beauty salons, road workers, etc. Those who need to protect themselves from UV rays must wear clothing that guarantees certain levels of protection from ultraviolet rays (Fig. 3.14).

The anti-UV fabrics are fabrics that offer a certain protection from the ultraviolet rays of the sun, measurable according to the UPF parameter (ultraviolet protection factor). This factor is an estimate of the capacity of a fabric to shield the UV radiation, see Eq. (3.1).

[26]https://ec.europa.eu/clima/policies/ozone_en

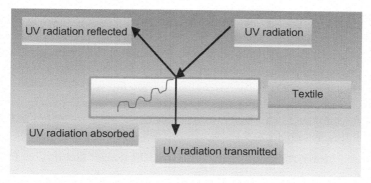

Figure 3.14 Possible interactions light textile.

$$\text{UPF} = \frac{ED}{ED_m} = \sum_{290\,\text{nm}}^{400\,\text{nm}} E_\lambda S_\lambda \Delta_\lambda \left/ \sum_{290\,\text{nm}}^{400\,\text{nm}} E_\lambda S_\lambda T_\lambda \Delta_\lambda \right. \qquad (3.1)$$

where

E_λ = erythemal spectral effectiveness

S_λ = solar spectral irradiance

T_λ = spectral transmittance of fabric

Δ_λ = the bandwidth in nm

λ = the wavelength in nm

For example, a fabric classified as UPF 50 allows only it allows only 1/50 of the UV rays to pass through it, that is, it blocks 98% (49/50) of UV radiation. The UPF is measured in the laboratory using a spectrophotometer together with an artificial light source, with a method such as the Diffey Robson method [21] for the instrumental measurement of SPF (sun protection factor).

It varies from fabric to fabric and can have values between 5 and 80, and anyway all fabrics offer protection from UV rays. Fabric construction factors are the most important for radiation shielding, it includes weave, weave density, cover factor, porosity, weight, and thickness. All the factors are interrelated and influence each other.

The thicker the weave of the fabric, the higher the UV radiation protection because the fibers of tightly woven fabrics are closer together, then less UV radiation can pass through to the skin. For the woven fabrics of same weight, the plain weave designs give the highest protection. Knitted or woven fabrics alter protection due to interlacing, the open spaces where yarns cross. Stretched areas in a garment pull at the interlacing, permitting UV to penetrate. Woven fabrics are more UV protective than knitted fabrics.

Cover factor is defined as the percentage area occupied by warp and weft yarns in a given fabric area. Cover factor is related to UPF as follows in Eq. (3.2).

$$\%UV \text{ radiation transmission} = 100 - \text{cover factor} \qquad (3.2)$$

Porosity, that is, the number of pores per unit of fabric surface, has an inverse relation with cover factor. Therefore, with a decrease in porosity the UPF increases. The relative order of importance for the UV protection is given by: % cover factor > fiber type > fabric thickness. Consequently, fabric with the maximum number of yarns in warp and weft will give high UPF value.

We report below some UPF test standard for textiles and in Table 3.2 an UPF classification system.

- Australian/New Zealand Standard AS/NZS 4399:1996 "Sun protective clothing—Evaluation and classification" First UPF measurement/classification standard; Solar spectrum of Melbourne, Australia; Measurement in new, dry and unstretched state; UPF classification system (Table 3.2)
- European Standard EN 13758-1 "Textiles. Solar UV protective properties. Part 1: Method of test for apparel fabrics" It refers only to apparel fabrics; Measurement in new, dry and unstretched state; same measurement conditions of AS/NZS 4399, but solar spectrum of Albuquerque; Part 2 specifies the requirements for marking clothing if they respect strict requirements (e.g., UPF > 40 + some design requirements)
- USA Standard AATCC Test Method 183–2014 "Transmittance or Blocking of Erythemal Weighted Ultraviolet Radiation through Fabrics" UPF measurement according Standard AS/NZS 4399/96 (Solar spectrum of Melbourne, Australia); Measurement under practical everyday use conditions: new condition, wearing stress (stretching), wetting, mechanical stress (abrading), textile care (washing), artificial weathering for shading textile; the lowest determined UPF can be certified;

the textile can be labelled (UV STD 801 label); issued only by one of the UV Standard 801 Institutes.

Table 3.2   UPF classification system

UPF range	UV protection category	Effective UV transmission (T%)	UPF ratings
15 to 24	Good protection	6.7 to 4.2	15, 20
25 to 39	Very good protection	4.1 to 2.6	25, 30, 35
40 to 50, 50 and over	Excellent protection	≤ 2.5	40, 45, 50, 50+

The absorption band of all dyes extends to UV radiation spectrum (280–400 nm) and all the dyes, therefore, can act as UV absorbers. The extinction coefficients of these dyes determine their ability to increase the UPF of fabric. With the same fabric, dark colors offer better UV protection than light pastel colors thanks to the increase in UV absorption. the degree of UV protection may also vary due to the specific characteristics of the dye. Some of the direct, reactive and vat dyes substantially increase the UPF of 50+.

The UPF of dyes extracted from various natural resources [22] is within the range of 15–45 depending on the mordant used. Optical brightening agents (OBA) are used at the finishing operations to enhance the whiteness of textiles by UV excitation and visible blue emission. OBA can improve the UPF of cotton and cotton blends, but not of fabrics that are 100% polyester or nylon. Limitation of OBA is that they mostly absorb in the UVA part of the day light spectrum but have a weak absorption in UV absorption around 308 nm which plays an important role in skin disease [23].

In conclusion, a dark, wrought, and heavy fabric has an anti-UV protection factor greater than a light, sparse, and light fabric. Rather heavy fabrics, even if light, can already have a high UPF. For clear and light fabrics, the situation instead becomes critical, because the UPF will easily be low. In these cases, it is possible to intervene by increasing the UPF with a specific anti-UV treatment.

A fabric that has low UPF, for example, 5, 10, or 15 requires adequate treatment that can generate significant increases in UV protection, opening new possibilities for the use of the article. The anti-UV treatments are important specially when we wear spring/ summer clothes characterized by light colour, light textures, light materials, that are not able to protect our skin from UV rays. Under the sun, nobody wants to wear a dark and heavy protective fabric that is certainly protective. The advantage of a treatment and the innovative aspect lies in obtaining good protection with a fresh and clear garment.

3.3.3 Water, Oil, and Dirt Repellent

One of the first applications of textiles was the protection of man from inclemency of weather (cold, hot, wind, rain, and snow), so we can understand the reason for many studies on hydrophobic and in general repellent treatments. Not only for umbrellas and therefore for water but also for all types of oils and dirt, therefore suitable for carpets, curtains, tablecloths, and coverings. Even in the medical field, hydrophobic tissues are very important, ensuring that bacterial growth, particularly active in damp environments, is hindered.

A fabric may be water-resistant if it is able to resist the penetration of water to some degree but not entirely. It is water-repellent if it is not easily penetrated by water, especially as a result of being treated for such a purpose with a surface coating. Finally, it is waterproof if it is impenetrable to water.

Ingress Protection Rating scale (or IP Code) is a scale from 0–8 that classifies the degrees of protection provided against the intrusion of solid objects (including body parts such as hands and fingers), dust, accidental contact, and water in electrical boxes. A clothing with IP equal to 8 is a fabric that protects against the effect of immersion in water under pressure for long period.

For conventional textile finishing, water and oil repellent properties are achieved using wet-chemical treatment with perfluorinated organic compounds. Good oil-repellency requires particularly long fluorocarbon chains. However, molecular fragments of the finishing chemicals may be released during the original treatment as well as the washing and re-impregnation stages. These fragments or their reaction products include perfluorooctanoic acid and perfluorohexanoic acid. These compounds are toxic environmental pollutants that are bioaccumulative and suspected of being carcinogenic.

It is therefore necessary to establish more efficient and more environmentally friendly finishing processes that release fewer pollutants and avoid fluorocarbon treatments as far as possible.

The application to surface treatment materials is of great interest in the industrial field. The development of these treatments results from a high level of research progress, an increase in their performance and potential and a reduction in costs; for this reason, in recent years, these techniques have been optimized and have made it possible to improve the surface properties of conventional polymers or other types of materials, affecting many industrial sectors, from mechanical applications to biomedical.

Plasma is an ionized or partially ionized gas. The present species that constitute it can usually be divided into neutral atoms and molecules, negative and positive ions, radicals, electrons, and photons. The interaction between these species and the surface of the treated material activates in turn processes of etching, grafting, activation, and deposition of surface films. The plasma treatment is therefore a surface modification process that is applied to materials to modify their properties. The hot plasmas are applied for the modification of the surface properties of materials that are able to withstand high temperatures. On the other hand, cold plasmas allow the treatment of substrates even with a low melting point such as polyethylene and polypropylene. It is important to underline that the modifications introduced by a plasma treatment involve only the superficial layers of the substrate and do not alter the general physical-mechanical properties of the material: in a textile, for example, the properties of breathability and hand are not altered.

The fundamental processes that occur during the treatment of a material are: insertion reactions of atoms or whole chemical groups (grafting), generation of free radicals on the surface (surface activation), deposition of polymers in the form of thin layers adherent to the surface (film deposition), and phenomena of surface ablation of the material (etching).

Plasma treatments can be carried out under vacuum or at atmospheric pressure. The vacuum systems are the first to have given results and to have been used at an industrial level in the textile sector; however, they have some disadvantages: among the main ones, a high degree of cleaning of the machine to limit contamination during treatments, the non-applicability to continuous processes and therefore a certain precaution in manipulating the tissues at the end of the treatments themselves to avoid deactivation phenomena. Because of all this, over the last few years atmospheric pressure plasma technology has become increasingly widespread, which involves the use of machines able to work continuously, fairly flexible and easily positioned at the beginning of the process. The atmospheric plasma technology for fibers and yarns considerably increases the wettability of the fabrics, favoring a stable adhesion over time of dyes without organic solvent. It allows to confer the long-term solidity of the colors in the printing and dyeing processes, together with a reduction in the raw materials used in conventional processes.

Plasma processes make it possible to reduce the use of chemical products even in the application of resins in solution or aqueous dispersion, guaranteeing high quality textile products, as well as being able to impart new properties to the fibers without the aid of subsequent coatings. The plasma treatment takes place in roll-to-roll mode and can be easily integrated in line with the production process. The technology has already been successfully applied to fabrics in natural and synthetic fibers to obtain properties such as hydro and oil-repellency, anti-pilling, easy-care, antibacterial, and increased dyeability.

Photocatalysis is the phenomenon in which the substance is, the photocatalyst, modifies the speed of a chemical reaction through the action of light. This phenomenon was discovered in 1972 by Fujishima and Honda [24]. Photocatalysis uses solar energy to make molecules active which, when illuminated by light with an appropriate wavelength, induce the formation of strongly oxidizing reagents capable of decomposing the organic and inorganic substances present. Photocatalysis is, in essence, an accelerator of oxidation processes already active in nature. The substances that modify the speed of a chemical reaction, through the action of light, are semiconductors. When a semiconductor is hit by photons having hv energy greater than its E_{gap}, an electron (e^-) is able to migrate from the valence band (VB) to the conduction band (CB), generating a vacancy electronic (h^+) or hole at the upper limit of the valence band. The components of the photogenerated pair are able, respectively, to reduce and oxidize a substance adsorbed on the surface of the photocatalyst (see Fig. 3.15). If the semiconductor is in contact with H_2O, the hole can produce hydroxyl radicals ($OH^\bullet$), while the electrons are enough reducing agents to produce from the oxygen the superoxide anion (O_2^-). These two highly reactive species are able to decompose the adsorbed substances. A known example is titanium dioxide, TiO_2, which exists in three crystalline forms: anatase, rutile, and brookite. High photocatalytic activity is mainly achieved by anatase. In the microbiological field several studies have been carried out to determine the antibacterial activity of the photo-excited TiO_2 particles with UV lamps [25]. In addition to antibacterial activity, we also mention self-cleaning and its use in products such as paints.

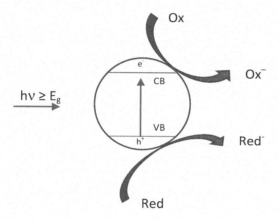

Figure 3.15 Scheme of photocatalysis reactions.

3.3.4 Antimicrobial Coatings

Bacterial infections are among the leading causes of death in the world. A simple gesture like washing hands changed the face of medicine. It was a Hungarian doctor who, in the mid-19th century, saw women who had recently died of sepsis to understand the cause. Ignaz Semmelweis (1818–1865) worked at a hospital in Vienna in obstetrics and noticed that in a pavilion, run by doctors, many women died after giving birth to sepsis or puerperal fever (about 11%) while in another pavilion, where helping women to give birth was only midwives, deaths were just 1% [26]. The answer came to Semmelweis from the autopsy on a man, a dear friend and colleague of his, who died after a brief illness: in his body he found the same injuries that were found in the bodies of the new mother that the doctors of the hospital dissected as research and normal practice. A few days earlier, Ignaz recalled, his friend had injured himself while performing an autopsy on a post-natal woman. The infection occurred by contact, in the hospital doctors and students went directly to the delivery rooms after performing autopsies and no one thought they had to wash their hands. It was precisely with the hands infected by the dissections performed on the rash deaths that the gynaecologists spread the contagion, for this reason the pavilion conducted by the midwives was healthier. To verify his thesis, the Hungarian doctor agreed with his colleagues and students to disinfect their hands with calcium chloride before entering the delivery room. The decline in deaths from sepsis was a real collapse: it was 1847 and in one year the pavilion of obstetric physicians also stood at 1% of deaths. It took forty years and work of Louis Pasteur (1822–1895) on bacterial contamination because the intuition of Ignaz Semmelweis has been accepted and applied by everyone. In 1894, the great doctor was able to have a worthy funeral monument erected from the birthplace of Budapest.

Materials with antimicrobial function are the first defense against a bacterial attack, we can divide them into two classes depending on how this action is performed (Fig. 3.16).

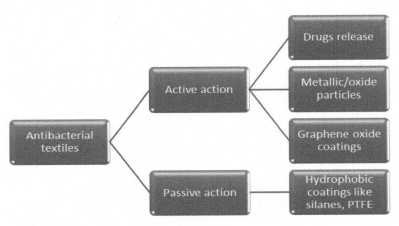

Figure 3.16 Types of antibacterial action.

We are talking about active action when drugs or specific metal particles or metal oxides in a humid environment release substance that can kill or inhibit bacterial growth. In the case of passive action, the bacteria do not grow on the material, but can proliferate around.

When a microorganism colonizes a new environment, its way of multiplying is not constant but depends on the characteristics of the environment, on the temperature and on the type of microorganism. The temperature greatly influences the speed of chemical reactions, metabolism, and consequently also the growth of microorganisms. Gradually increasing the temperature, the reaction speed increases until it reaches a maximum value. By raising the temperature further, the speed decreases dramatically because at high temperatures enzymes, transport proteins, and other proteins denature, and the cell membrane is destroyed. Water is essential for bacteria; a moist environment is therefore conducive to their growth. The type of fabric favors or not bacterial growth. Polyester, for example, due to its hydrophobic characteristics, is not able to absorb sweat as it happens with cotton, which is why a polyester fabric stink much more.

The antibacterial products also protect the fabric from degradation, reduce the formation of odors following sweating, protect against the transfer and spread of bacteria. The main applications are in medical field, but also in sportswear and military clothing.

The resistance of many bacteria due to the excessive use of antibiotics has led to an increase in the demand for antibacterial clothing and complements in hospitals, nursing homes, and kindergartens. In the field of burn treatment there are applications of antibacterial textiles, as far as bandages and plasters are concerned, able to support the healing process.

In the case of diabetic patients, special socks prevent the spread of infections. One of the effects of diabetes is the so-called diabetic foot. Diabetic disease causes narrowing of the arteries leading to reduced blood supply to the lower limbs, particularly to the feet. The possible progressive dysfunction of peripheral nerves causes loss of sensitivity in the lower limbs, a condition that leads to neglecting wounds and infections. The conclusion may be amputation, but this can be prevented by observing and periodically treating the lower extremities.

There are on the market products as X-static[27] silver-coated nylon, and products as chitosan[28] based (Fig. 3.17).

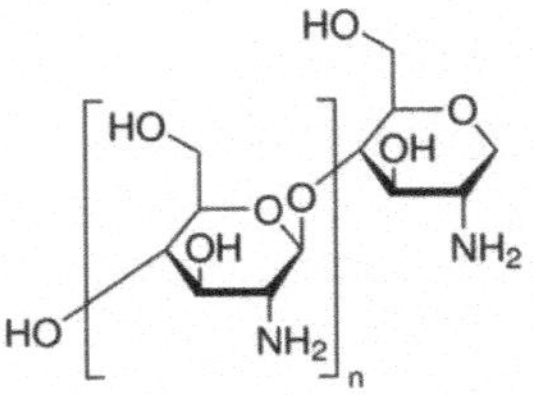

Figure 3.17 Chitosan structure.

What is required of all these materials, however, is to have high efficacy with a low degree of toxicity.

There are photoactive coatings that destroy microbes thanks to an oxidizing process during exposure to light (see TiO_2).

There are also controlled release systems of biocides. The antimicrobial agent is immersed in a usual matrix and the release occurs when the device comes into contact with liquid media. Due to the special interest in coatings to prevent biocontamination, different ways have been investigated to prepare coatings with biocidal properties. It is possible to incapsulate biocides in cyclodextrins [27], in inorganic sol–gel matrix [28], and antimicrobial polymers [29]. Few examples of antibacterial agents (see Fig. 3.18). Triclosan have been banned in US from FDA.[29]

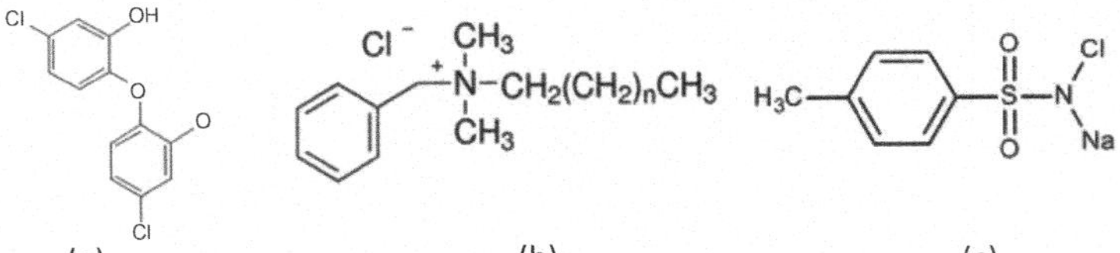

(a) (b) (c)

Figure 3.18 (a) Triclosan, (b) benzalkonium chloride, and (c) cloramine T.

Recently a different kind of antibacterial agents has been testate: the family of graphene [30], a highly studied material for its very

[27]http://noblebiomaterials.com/xstatic-textiles/

[28]https://www.swicofil.com/

[29]https://www.fda.gov/news-events/press-announcements/fda-issues-final-rule-safety-and-effectiveness-antibacterial-soaps

good properties in thermal and electrical conductivity, has given excellent results also to bacteria [31].

3.4 Method to Functionalize AgNPs to Materials

Functionalizing a material means modifying its surface composition by incorporating elements or functional groups that change surface properties. The process allows the material itself to perform specific and previously designed functions, which would otherwise not be obtained from the chemical-physical characteristics of the material itself.

The functionalization process is the result of the use of innovative technologies to modify the macrostructure and/or the composition of the material within the production process or the adoption of specific treatments on the finished product that modify the chemical-physical behavior of the material or of its surface layer. There are very interesting perspectives in the use of technology and in the possibility of giving functionalities to clothing. The growing interest in this field shows that there are opportunities to increase the added value and therefore the profits of the entire textile chain. The implementation of new technologies requires close cooperation between the various parties involved: chemical industries, fiber producers, weavers, designers and fashion houses, activewear clothing manufacturers, etc.

The strategies for antimicrobial finishing in clothing textiles have been imported by the medical sector, they are developing more and more and find enormous possibilities of development also in the field of home textiles. The antibacterial substances on the fiber prevent the formation of fungi and microorganisms, as well as of course the formation of bad smells.

The major problem concerning functionalization is stability, for example, to washes, over the year's methods have therefore been proposed to increase it. The use of binders, cross-linking agents, special polymers, and not least physical pre-treatments on the fibers, have been intensively studied.

Functionalizing a tissue with AgNPs mainly means giving it antibacterial properties. Many textiles have been used for this purpose, cotton has been the most explored, at least at the beginning.

In the deposition of AgNPs on a textile fiber there can be electrostatic interaction between positive silver ions, before reduction, and any negative charges of the material, or the nanoparticles can diffuse in the fiber, or there may be an interaction between silver and atoms such as sulfur, nitrogen, oxygen, in this case the chemical composition of the textile (wool, silk, cotton, polyamide...) plays a very important role. The interaction between silver and vat[30] dyes are still considered.

We can have three different application methods: (1) dipping, (2) sonochemical, and (3) layer-by-layer (Fig. 3.19).

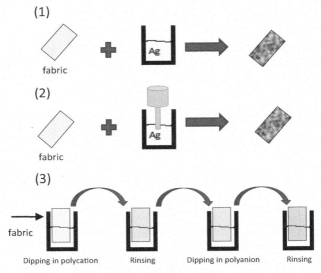

Figure 3.19 Application methods (1) dipping, (2) sonochemical, and (3) layer-by-layer.

---

[30]The vat dyes, whose progenitor is indigo, are insoluble, but in a strongly reducing environment they are reduced to give the so-called "leuco" form, soluble and colourless. By soaking the fiber of the "leuco" form solution and then leaving it exposed to air or possibly treating it with suitable oxidants, the "leuco" form restores the dye, which precipitates inside the fiber thus remaining permanently fixed there. The name of vat, which continues to be used to indicate the device in which the "leuco" form dye is reduced, derives from the ancient indigo dyeing process: this was introduced into a vat in which organic materials were left to rot. different, thus creating a strongly reducing environment with putrefaction. This primitive reduction method has been replaced by those using chemical reducers, such as sodium hydrosulphite. Vat dyes provide dyes which are generally less brilliant than those obtained, for example, with azoic or triphenylmethane dyes, but much more stable.

3.4.1 Dipping

Immersion treatment is the most common method to attach AgNPs to fiber. We generally have three steps: material preparation, treatment, and completion. Prior to dipping, it is very important that material is clean and then it is washed to remove contaminants and improve the adhesion between the fabric and the coated film. Then the textile is dipped in AgNPs or in a solution of silver salts, the time of treatment can vary from minutes to hours, depending on kind of material [32]. At this moment the use of binders is often recommended, among the highly used binders there are the acrylic resins and copolymers, vinyl acrylic resins, PVC Resins, vinyl acetate resins, EVA, styrene butadiene, acrylonitrile butadiene, polychloroprene, polyvinyl alcohol, carboxyl methyl cellulose (CMC), DMF polyurethanes, water polyurethanes, natural rubber, starch, resins phenolic, melamine, and ureic thermosets [33].

The curing is the final step, normally it is carried out at high temperature to dry the material and to fixing better the treatment.

3.4.2 Sonochemical

The discovery of ultrasound dates to the end of the 18th century and is due to Italian Scientist Spallanzani (1729–1779) who, observing the flight of the bat, noticed his ability, despite being almost blind, to avoid the obstacles that stand in the way of the flight and its ability to catch small insects. He hypothesized that this fact happened thanks to the emission of sound waves [34]. The emission of sound waves was confirmed later, with modern diagnostic methods by the American physicist Griffin of the University of Harvard who, repeating the Spallanzani experience with modern means, succeeded in demonstrating the existence of sound waves emitted by bats, sound waves not perceptible to the human ear [35]. With these means they could see that the bat emits cries of the frequency of 45,000 periods per second and of the duration from 5/1000 to 1/100 of a second; these cries when they encounter an obstacle, are sent back and the echo, perceived by the ear of the bat, allows to change the trajectory of the flight, thus avoiding the collision. The first practical applications of ultrasonic waves date back to the beginning of the last century. After the sinking of the Titanic (1912), the naval technicians set out to devise efficient methods for quickly signaling obstacles on the route of the boats at sea, useful were the ultrasonic waves that are transmitted in the water. The outbreak of World War I accelerated the process of building ultrasound equipment based on the studies of a French physicist Langevin (1872–1946). This built an ultrasonic vibrator capable of transmitting sound messages in South of France [36]. In 1927, the first works appeared in the United States concerning the characteristics, mechanical, thermal, chemical, and biological effects of ultrasound. During the World War II, to exploit its war applications, it was the subject of many studies by German researchers and allies; the remarkable development of scientific investigations, gave a remarkable impulse to the therapeutic applications of ultrasounds. In 1949, after the end of the war, the 1st First International Congress on Ultrasound in Medicine was held in Erlangen during which German and allied researchers could compare their experiences. It was precisely on this occasion that ultrasound research became a branch of medicine.

The ultrasounds are elastic waves in a material medium with a frequency higher than the audibility limit, conventionally set at 20 kHz It was one of the first cases in which the piezoelectric effect was used for their generation. This discovery has laid the foundations for a great spread of their use. Piezoelectric materials have the characteristic of lengthening or shortening as a function of the tension to which they are subjected. This makes it possible to transform an oscillating voltage signal into a mechanical vibration of equal frequency. Intensity and frequency of mechanical vibration thus depend on the intensity and frequency of the electrical signal. If the frequency of the electrical signal is higher than 16 kHz, then an ultra sound wave is produced. The main technological applications of ultrasound are: sensors (proximity sensors, structural diagnostics); industrial washing (ultrasonic washing baths); assembly and / or processing of plastic materials (welding of metals, machining with chip removal).

Sonochemistry studies the chemical reactions that take place in a solution radiated by ultrasounds [37]. This irradiation gives rise, due to an intensity of the field greater than a certain threshold value, to a cavitation phenomenon in the solution, that is to the formation, growth, and implosive collapse of bubbles in a liquid. The gaseous microcavities (bubbles) present in the solution, subjected to successive expansions and contractions induced by the oscillating field of sound pressure, are enlarged and then implode, producing areas of very high temperature and pressure. In fact, cavitation collapse produces intense local heating (~5000 K), high pressures (~1000 atm) and enormous heating and cooling rates (>109 K/sec) and liquid flows (~400 km/h). In these extreme conditions chemical reactions of considerable interest can occur in the synthesis of organic substances, polymerization processes, degradation of toxic and harmful substances. With the application of sonochemistry techniques it is also possible to obtain amorphous materials which, outside the extreme conditions typical of sonochemistry, would naturally tend to crystallize.

The ultrasound-assisted procedure can be used as a progressive technique for the deposition of nanoscale metals and metal oxides on various materials. The advantage of this method is the absence of stabilizing surfactants during the process and the product obtained is relatively pure with particle sizes very small.

3.4.3 Layer-by-Layer Deposition

The layer-by-layer (LbL) assembly of multilayered polymer is a technique that allows nanoscale control over sample thickness and is extensively applicable in packaging, optical films, and coating planar and particulate substrates.

This method is very simple to apply, born from the idea spread in antiquity and revived in 1774 from American scientist Franklin [38] that the oil scattered on the surface of the sea calmed the waves. We have to get to 1920 when American scientist Langmuir transferred a layer of oil on a solid substrate and in 1935 American physicist Blodgett transferred more layers to a solid substrate. In 1966 Iler originally proposed the LbL technology and prepared multilayered films through alternating deposition of colloidal particles of opposite charges [39]. The method was modified and improved over the following years. Therefore, the LbL technology was extended from two-dimensional to three-dimensional space so that its research and application could be greatly expanded. Among different deposition techniques, the LbL method has focused the attention of many research groups.

LbL requires only a minimal surface charge than this can be induced in different substrates such as glass, quartz, silicon wafers, and mica. A charged substrate is immersed in a solution of an oppositely charged colloid to adsorb the first monolayer, then a washing cycle follows to remove unbound material and preclude contamination of the subsequent oppositely charged colloid, finally the coated substrate is submerged to deposit a second layer and the multilayered structure is formed. Some LbL processes require no washing cycles thus shortens the duration of the assembly process.

References

1. South Tyrol Museum of Archaeology, Bolzano, Italy.

2. (a) Prinoth-Fornwagnera, R., Niklausb, Th.R. *Nuclear Instruments and Methods in Physics Research Section B: Beam Interactions with Materials and Atoms* (1994) **92**, pp. 282–290. (b) Kutschera, W., Rom, W., *Nuclear Instruments and Methods in Physics Research Section B: Beam Interactions with Materials and Atoms* (2000) **164-165**, pp. 12–22.

3. Snodgrass, M.E., *World Clothing and Fashion: An Encyclopedia of History, Culture, and Social Influence* (2013), Routledge.

4. Homerus, Odissey XIX; XVIII, 292–294.

5. Berzelius, J.J., *Annals of Philosophy* (1814) Vol. III.

6. Guareschi, I., Storia della Chimica XI, Jöns Jacob Berzelius e la sua opera scientifica (1915) Unione Tipografico Editrice Torinese, Torino.

7. Barber, E.J.W. *Prehistoric Textiles* (1991), Princeton University Press, New Jersey, U.K.

8. Frei, K.M., Mannering, U., Kristiansen, K., Allentoft, M.E., et al. *Scientific Reports* (2015) **5**, pp.10431–10438.

9. Ryder, M.L. *Sheep and Man* (1983), Duckworth, London.

10. Kuhn, D. Tracing a Chinese legend: In search of the identity of the "first sericulturalist," *T'oung Pao*, Second Series, **70** (4-5) (1984) pp. 213–245.

11. Richthofen, F., *Tagebücher ans China* (1907), D. Reimer (E. Vohsen), Berlin.

12. Andrews, C. *Egyptian Mummies* (1984), Harvard University Press.

13. Komuraiah, A., Kumar, N.S., and Prasad, B.D. *Mech Compos Mater* (2014) **50**, pp. 359–376.

14. Bredemann, G., Garber, K. Die grosse Brennessel Urtica dioica L. Forschung über ihren Anbau zur Fasergewinnung (1959), Akademieverlag, Berlin.

15. Lucio Anneo Seneca Epistulae morales ad Lucilium book XIV 90 20.

16. Leonardo da Vinci Il Codice Atlantico foglio 985 recto 1979 Hoepli Firenze, Biblioteca Ambrosiana, Milano, Italy.

17. Chardonnet Comte Hilaire de Patent GB189324638 (A) *Improvements in the Manufacture of Artificial Silk* (1894).

18. Hooke, R., *Micrographia: or, Some physiological descriptions of minute bodies made by magnifying glasses. With observations and inquiries thereupon* (1665), John Martyn and James Allestry, London.

19. De Réaumur, R-A.F. Mémoires pour servir à l'histoire des insectes (1749).

20. (a) Carothers, W.H., *Chem. Rev.* (1931) **8** (3), pp. 353–426. (b) *The First Nylon Plant: A National Historic Chemical Landmark* (1995), ACS.

21. Diffey, B., and Robson, J. *Journal of the Society Cosmetic Chemistry* (1989) **40**, pp. 127–133.

22. Gupta, D. *Colourage* (2007) **54**, pp. 75–80.

23. (a) Grifoni, D., Bacci, L., Di Lonardo, S., Pinelli, P., et al. *Dyes and Pigments* (2014) **105**, pp. 89–96. (b) Grifoni, D., Bacci, L., Zipoli, G., Albanese, L., and Sabatini, F. *Dyes and Pigments* (2011) **91**, pp. 279–285. (c) Hustvedt, G., and Cox Crews, P. *Journal of Cotton Science* (2005) **9**, pp. 47–55.

24. Fujishima, A., and Honda, K. *Nature* (1972) **238**, pp. 37–38.

25. Gupta, K., Singh, R.P., Pandey, A., and Pandey, A., *Beilstein Journal of Nanotechnology* (2013) **4**, pp. 345–351.

26. Céline, L. *Semmelweis* (1951) Atlas Press (GB).

27. (a) Gawish, S.M., Ramadan, A.M., Mosleh, S., Morcellet, M., Martel, B. *Journal of Applied Polymer Science* (2006) **99**, 2586–2593. (b) Radu, C.-D., Parteni, O., Ochiuz, L. *Journal of Controlled Release* (2016) **224**, pp. 146–157.

28. Mahltig, B., Haufe, H., and Böttcher, H. *Journal of Material Chemistry* (2005) **15**, pp. 4385–4398.

29. Kenawy, E.R., Worley, S.D., and Broughton, R. *Biomacromolecules* (2007) **8**, pp. 1359–1384.

30. Geim, A.K., and Novoselov, K.S. *Nature Materials* (2007) **6**, pp. 183–191.

31. Krishnamoorthy, K., Navaneethaiyer, U., Mohan, R., Lee, J., and Kim, S. *Applied Nanoscience* (2012) **2**, pp. 119–126.

32. (a) Chenga, D., Hea, M., Rana, J., Caia, G., Wua, J., and Wang, X., *Sensors & Actuators: B. Chemical* (2018) **270**, pp. 508–517. (b) El-Rafie, M.H., Ahmedb, H.B., and Zahranba, M.K. *Carbohydrate Polymers* (2014) **107**, pp. 174–181.

33. (a) Bonet Aracila, M.A., Bou-Beldaa, E., Monllora, P., and Gisbertb, J. *The Journal of the Textile Institute* (2016) **107**, pp. 300–306. (b) Asaduzzaman, Kamruzzaman, Hossein, F., Miah, M.R., and Sultana, Z. *International Journal of Scientific & Engineering Research* (2016) **7**, pp. 710–716.

34. Spallanzani, L., "Lettera sopra il sospetto di un nuovo senso nei pipistrelli" Giornale fisico medico, Pavia 1794, Vol. I, pp. 197–244.

35. Griffin, D. *Echoes of Bats and Men* (1960), Heinemann.

36. Lewiner, J. *Japanese Journal of Applied Physics* (1991) **30**, pp. 5–11.

37. (a) Perelshtein, I., Applerot, G., Perkas, N., Guibert, G., Mikhailov, S., and Gedanken, A. *Nanotechnology* (2008) **19**, pp. 245705–245711. (b) Perkas, N., Amirian, G., Dubinsky, S., Gazit, S., and Gedanken, A. *Journal of Applied Polymer Science* (2007) **104**, pp. 1423–1430.

38. Mertens, J., *Physics Today* (2006) **59**, pp. 36–41.

39. Iler, R.K., *Journal of Colloid and Interface Science* (1966) **21**, pp. 569–594.

Chapter 4

In situ Synthesis of Silver Nanoparticles

4.1   A Long Time Ago

My research on the application of silver nanoparticles on textiles began in 2007 as part of a project involving universities and companies from the Lombardy (a big region of Italy).

The aim of the project was to stimulate a technological leap in the textile sector "made in Italy" using the control of matter at the molecular and nanoscale techniques.

The project involved a joint work of basic research and industrial feasibility studies to connect SMEs and large companies in the textile sector in Lombardy along with public research laboratories of the most important in Italy. The objective was to exploit the enormous potential offered by nanotechnology, with the aim of imparting innovative and unique features in fibers and synthetic and natural fabrics.

The treatment of the yarn in order to increase comfort, fit, and aesthetics is one of the most important issues in the clothing field. Unfortunately, the transfer of these properties is often linked by little lasting treatments over time and limited effectiveness against subsequent washing or prolonged exposure to solar radiation.

The project included two different types of processes, the first one was the treatment of a traditional fabric or wire by nanoparticles according to an appropriate procedure to be developed, and the

Silver Nanoparticles: Synthesis, Properties, and Applications
Anna Facibeni
Copyright © 2023 Jenny Stanford Publishing Pte. Ltd.
ISBN 978-981-4968-21-8 (Hardcover), 978-1-003-27895-5 (eBook)
www.jennystanford.com

second was the insertion of nanoparticles directly into the fiber during the spinning in order to make the most effective and lasting cohesion.

The first road, certainly much less onerous from the point of view of the financial investments and project timelines, however, presents as a significant disadvantage, at least in the configuration of current industrial technologies, the poor permanence of the treatments on apparel. Instead, the functionalization of the fibers themselves could be extremely stable, but very complex to achieve and difficult to control, mainly if we do not want to induce chemical/physical modifications and structural changes to the finished product. Wanting to keep open the two options functionalization classically pursued on an industrial level, investigations have been done on the two strands in parallel.

The research also included nanoparticles of gold, copper, and oxides of transition metals.

Gold, presenting considerable activity hydrophobic and a strong stability of the nanoparticle, would be a prime candidate to replace current treatments in highly volatile trade and little wear resistant [1].

Regarding the copper appears clear how a possible modification in catalytic activity and oxidation-reduction of the Cu-nanometer could allow a reduction of industrial costs, with huge implications in the field [2].

Concerning the oxides essentially, we can speak of TiO_2 and ZnO, widely used in the industrial field [3, 4]. At the end of the book we will see the application of our methodology for these two oxides.

The existing fabric processing techniques by means of microparticles show in addition to the problem related to the ratio of amount of material used and its chemical activity, also problems of resistance of the treatment to washing and wear. The microparticles in fact have a poor adhesion to the fabric and tend, after a low number of washes, to disperse. The nanoparticles, due to their extremely small size, instead show a greater adhesion to the fiber.

Many products on the market take advantage of the co-extrusion of active particles in the polymer fibers, but the problems related to the large dimensions of the Ag particles with respect to the inserted fiber sections, contribute, with the cost of the material, to increase the cost of the finished product. The extremely compact size of the latter would reduce the fiber/particle cohesion problems and to treat the "whole object as a" unique active surface.

At the beginning of the project, it was decided to deposit the silver nanoparticles by a physical technique such as pulsed laser deposition, used for many years in our laboratory. As discussed in Chapter 2, the PLD is a thin-film synthesis technique that uses the laser ablation process.

In our laboratory we use an excimer laser, with a wavelength in the UV wave, the range 10–20 ns pulse width, repetition frequency of impulse up to 200 Hz, energy density deposited on the target up to 500 mJ cm^{-2}. The deposition chamber is in the ultra-high vacuum (UHV), with a sample holder equipped with a side of the handling and control of the substrate temperature.

The advantage is to produce pure materials the disadvantage is the difficulty to apply that method to industrial production.

For this reason, it attempts a chemical treatment that would allow more consistent to reduce production costs and times. The first step of our research has been to examine and try different types of reductant.

I began to look at what had been done in the literature.

In the last years, the increasing demand of reduce generated hazardous waste has resulted in integration of industrial processes with green reagents.

The novel idea born in our group was introducing the material in the solution before the reduction, the word *in situ* is derived from the Latin (been in place in + ablative of situs, site) and means on the spot. This term is intended to indicate something that remains in the place. This expression is often used in medicine to indicate the physiological positions of body parts but also to indicate a particular disease that remains in its pathological headquarters, used in oncology to define a malignant tumor at its early stage, that is when it involves only the cells from which it was originated and has not spread to adjacent tissues. In the case of malignant tumors of epithelial type, or carcinomas, this concept is expressed when the cancer cells have not yet passed the basement membrane of epithelial origin.

In engineering it refers to operations that are carried out in the same place where you get the work built or the final product. In the chemical, it may be reported to the reagents or catalysts which are prepared in the reaction (or in the chemical reactor).

We have used the colloidal synthesis suitably amended to achieve a good result.

4.2 Colloids

The name colloid comes from the Greek κόλλα. Originally indicated certain substances, such as Arabic gum, gelatine, starch, etc., which had in common the absolute lack of crystalline form and gave viscous solutions. It was first used by the English chemist T. Graham (1805–1869) in 1861, who with the Italian chemist F. Selmi (1817–1881) is considered the father of colloids chemistry.

The medical terminology even today is essentially of Greek origin, due to the enormous influence that Corpus Hippocraticum had on ancient medicine and this on medieval and modern medicine, from Greek derive the names of the phases of investigation and action of the doctor. The names of the categories of drugs are derived from Greek. The names of the branches of medicine are derived from Greek. The names of diseases are derived from the Greek. From the Greek they derive the terms that have entered the common language, but that originally had a different meaning from the present one. Finally, the names of the parts of the body, preserved in mostly compound terms, derive from Greek.

The Corpus Hippocraticum is a collection of about seventy works in ancient Greek that deal with various themes, among which medicine stands out, written over several centuries and aggregated together in an unspecified period. The attribution of the individual works is extremely complex: some are attributable to Hippocrates of Kos (approximately 460 a. C.–370 a. C), while others undoubtedly derive from the influence he had in later centuries.

We speak of colloidal system when a chemical species constitutes an extremely dispersed phase in a continuous phase. We can have fumes (solid dispersed in a gas), colloidal systems (solid dispersed in a liquid), emulsions (liquid in a liquid), and fogs (liquid in a gas).

When we talk about colloids, we mean a system composed of one or more phases dispersed in a continuous medium. We define a heterogeneous system as a mixture of two or more substances, when their composition varies locally. Consider heterogeneous biphasic systems in which one substance is uniformly dispersed in the other in the form of particles. In these systems we talk about the dispersed phase containing the particles and the dispersing phase the medium in which the substance is distributed. We can distinguish mixtures (large particles >0.1 µm), colloids (medium particles 0.1–0.001 µm), and solutions (small particles <0.001 µm) according to the size of the particles.

A colloidal dispersion with the naked eye seems identical to a solution but it is not so. Depending on the type of phase dispersed (liquid, solid, gas) we can have aerosols, foams, gels, etc.

4.2.1 Lyophilic and Lyophobic Colloids

The division of colloids into lyophobic and lyophilic is not a formal division but expresses a substantial difference between the two classes. For the lyophobic colloids, the colloidal dispersion state is an unstable state ($\Delta G > 0$), for the lyophilic ones it is instead a stable state ($\Delta G < 0$).[31]

Lyophilic colloids are like the solvent (lyo from Greek λύω, to loosen, dissolve and philo from Greek φιλο loving), are soluble and are stable systems, such as blood plasma. When the dispersing medium is water, we speak of hydrophiles.

In lyophilic colloids, the particles in the dispersed phase undergo intensive interaction with the molecules of the surrounding liquid. The particle surfaces are strongly solvated, and the specific free surface energy (surface tension) at the separation boundary is extremely low. The conditions necessary for the formation of lyophilic colloids are achieved at room temperature if the interphase (surface) tension does not exceed several hundredths dyne cm^{-1} (N m^{-1}).

Lyophilic colloids are formed by the spontaneous dispersion of large clumps of a solid or drops of a liquid into minute colloidal particles (micelles). Such colloids are thermodynamically stable and therefore do not disintegrate when kept under the conditions necessary for colloid formation. The lyophilic colloids include critical emulsions (that is, emulsions formed near the critical displacement

---

[31]ΔG represents free energy of Gibbs.

temperature for two interacting liquids), the colloidal dispersions of micellar surfactants (soaps, certain organic pigments, and dyes), and the aqueous dispersions of bentonite clay. The gel that forms because of coagulation in the lyophilic colloids can be dissolved again by addition of the solvent (reversible) while the lyophobes are irreversible systems since there is no possibility for them to return to solution.

Lyophobic colloids (from Greek Φόβος meaning fear) are unstable systems, in which the dispersed particles do not have affinity with the solvent, if the dispersing medium is water, we will talk about hydrophobes. These systems can result in coagulation by addition of small amounts of electrolytes or small temperature increases. In this case we will need to use emulsifiers to stabilize them. An example is milk, an emulsion of fatty acids, casein (a protein), and lactose (carbohydrates).

In lyophobic colloids, there is little interaction between the particles in the dispersed phase and the surrounding medium. The interphase tension in these systems reaches a relatively high level: not less than several tenths dyne cm^{-1} at room temperature. Because of an excess of free surface energy, lyophobic colloids are thermodynamically unstable; that is, they show a constant tendency to decomposition. The decomposition of a lyophobic colloid involves the coagulation or coalescence of the colloidal particles, accompanied by a decrease in the free energy of the system. The aggregation stability (the ability to resist particle consolidation) of any lyophobic colloid is temporary, since it is determined by the presence of a stabilizer, a substance, adsorbed at the surface of the particles (drops), that prevents agglutination (fusion). Typical examples of lyophobic colloids include the hydrosols and organosols of metals, oxides, and sulfides, and maximally dispersed (except critical) emulsions.

The main characteristics of colloidal systems can be attributed to the existence of an interface between dispersed phase and dispersing medium. Each particle has a well-defined surface, with a high surface development, which has properties such as adsorption, coalescence, electrical layer so it must be considered when studying the behavior of colloidal systems.

Now we see some colloids features:

❖ Optical properties as Faraday–Tyndall effect, the light diffusion through a colloidal solution

❖ Kinetic properties as Brownian motion. The Brownian motions consist in continuous collisions with the surrounding molecules of the dispersing medium. This motion contributes to the stability of colloidal dispersion in that it counteracts the action of gravity which tends to sediment the dispersed particles

❖ Electrical properties due to the double electrical layer around the nanoparticles, as electrophoresis. Electrophoresis is a phenomenon whereby electrically charged colloidal particles move because of an electric field and move toward the cathode if positively charged and toward the anode if negatively charged. The speed at which they move depends on their size and on their charge.

We consider a nanoparticle as an insoluble particle in a medium. In the liquid it will develop a surface charge because the surface groups are ionized as a function of pH or because ions are selectively adsorbed on the surface (OH^-).

We will talk about potential zeta, indicated with $p\zeta$, which is a measure of repulsive forces. Z potential is a critic parameter that determines nanoparticles stability or aggregation in a dispersion. A high Z potential gives more stability to colloidal system because the electrostatic repulsions generated prevent the aggregation of dispersed particles. When the potential is lowest attractive forces overcome the repulsions and therefore it is easier the occurrence of processes such as coagulation and flocculation.

Schematically in Fig. 4.1, we can observe how $p\zeta$ influences the stability of dispersion.

If the value of the zeta potential is high (negative or positive) it is electrostatic repulsion, if instead the value of the zeta potential is low (negative or positive) the attractive forces prevail and therefore the particles unite.

It is possible to modify the Z potential by modifying the ionic composition of the system, that is, varying the concentration of the electrolyte already present or the value of the electrolytes while maintaining a constant concentration. If the electrolyte has a higher value, the Z potential decreases and therefore the repulsive forces between the particles decrease. It is also possible to use ionic surfactants.

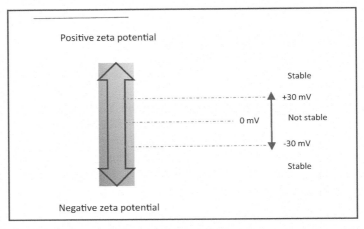

Figure 4.1 Influence of $p\zeta$ on dispersion stability.

When two or more colloidal particles collide, they can join (coagulation as a result of van der Waals forces attractions) in larger particles and get heavy until they precipitate. Therefore, the Brownian motion alone is not enough to ensure the stability of the suspension.

The destruction of a colloid can be obtained by heating or by the addition of an electrolyte, giving rise to coagulation or flocculation phenomena. Both phenomena lead to the formation of dispersed particles of greater mass and volume.

We referred to flocculation when several original particles form a cluster, the flocculum, thanks to interparticle bonds that affect the respective active sites, while maintaining their respective identity (flakes or protein denatured after heat treatment of the corresponding colloidal solution).

We talk about coalescence when in contrast several particles melt to form a larger mass and it is no longer possible to recognize the original particles (a liquid droplet that forms from smaller droplets).

The superficial tension is grandly modified only in lyophilic colloids and the viscosity of the dispersant increases only in lyophilic sol.

The method of synthesis can vary the surface charges of the silver nanoparticles, and this can also influence their antibacterial activity [5]. We can employ different capping agent to fabricate AgNPs with positive, negative, and neutral surface charges. The neutrally [6] and negatively [7] charged nanoparticles were synthesized according well know procedure, in the case of the positive charged we refer to a different procedure [8].

It is possible to determine the surface charge of the synthesized AgNPs by zeta potential analysis using a zeta potential analyser. Electrophoretic light scattering allows us to measure the electrophoretic mobility of particles suspended in a liquid, which is directly proportional to their zeta potential as described by Smoluchowski's formula [see Eq. (4.1)]. To measure the electrophoretic mobility of particles, an electric field is applied between the electrodes of the measuring cell with the sample and illuminated by a laser beam. The charged particles move toward the electrode of opposite sign, creating a frequency variation of the light scattered by the sample directly proportional to the electrophoretic mobility.

$$\zeta = \frac{4\pi\eta}{\varepsilon} \times U \times 300 \times 300 \times 1000 \qquad (4.1)$$

where ζ is zeta potential, η is the viscosity of solution, ε is the dielectric constant, U is the electrophoretic mobility.[32]

The use of coagulants to purify water dates to ancient Egypt [9].

4.3 Fast and Soft Reductant for Silver

Silver nanoparticles are metallic, then to bring the silver ion to metal is necessary the presence of a reducing agent.

One of the most widely used fast reducing is $NaBH_4$ [10], mainly used for the reduction of aldehydes and ketones. Silver reduction occurs according to the reaction [see Eq. (4.2)].

$$AgNO_3 + NaBH_4 \rightarrow Ag + \frac{1}{2}H_2 + \frac{1}{2}B_2H_6 + NaNO_3 \qquad (4.2)$$

---

[32]$U = v/V/L$, where v is the speed of particles (cm/sec), V is the voltage (volts), and L is the distance of the electrode.

This type of reduction produces nanoparticles of the order of 10–14 nm.

In this case the reductant is used in large excess also to stabilize the silver nanoparticles that are formed. The solution must be constantly stirred and maintained at a temperature of about 0°C. The use of this reductant would involve the use of some devices that do not easily agree with the SMEs, you must think about adapting an existing system more than a new design.

Reduction with hydrazine involve the same behavior, reductant is less expensive than $NaBH_4$, but its toxicity and hazard make them less attractive for us [11]. The molar ratio of silver nitrate/hydrazine was always calculated lower than 4:1 to avoid that to prevent the presence of unreduced silver (I) [see Eq. (4.3)].

$$4Ag^+ + N_2H_4 \rightarrow 4Ag^0 + N_2 + 4H^+ \qquad (4.3)$$

Furthermore, it should have skilled personnel, not always present in SMEs.

For this reason, we investigated others reductant, weaker, like ethanol, sugars, citrate, and ascorbic acid.

Ethanol is a reductant user friendly, the reaction is easy and fast, but this alcohol has the disadvantage of being flammable, and it is expensive [12].

Sugars, for example, fructose, [13] offer good results with low costs, we tried, but we wanted an alternative reducing agent.

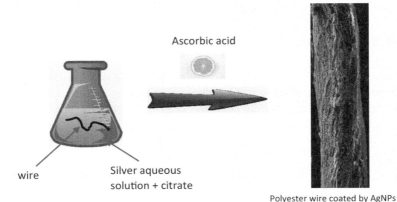

Ascorbic acid

wire

Silver aqueous
solution + citrate

Polyester wire coated by AgNPs

Figure 4.2 Scheme of synthesis *in situ*.

Schematically Fig. 4.2 shows the synthesis *in situ* with ascorbic acid (or citrate) as reductant.

4.3.1 The First One: Trisodium Citrate

4.3.1.1 Brief history

Sodium citrate is the salt of citric acid known for the Krebs cycle of fundamental importance in all cells that use oxygen in the cellular respiration process (see its structure in Fig. 4.3).

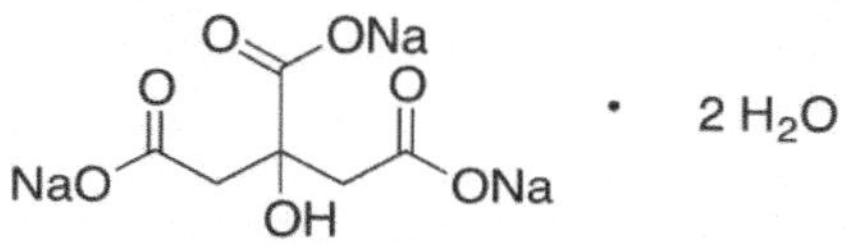

Figure 4.3   Trisodium citrate dihydrate.

The basic paper on citric acid is "The role of citric acid in intermediate metabolism in animal tissues" written by H. A. Krebs[33] and W. A. Johnson. The manuscript rejected by *Nature*, was published in 1937 by *Enzymologia* [14]. Today this review took the name of *Cellular and Molecular Biochemistry* (edited by Springer).

The first result that the authors report is that citrate has a catalytic effect on the breathing of striated muscle: the addition of citrate to freshly shredded tissue causes an increase in oxygen consumption greater than that necessary to oxidize the added citrate.

At room temperature, sodium citrate occurs in the form of a white solid, odorless and with a salty, slightly sour taste. Both its constituents (sodium and citric acid) are abundant in nature and within the human body the sodium taken through the diet is very important for the control of blood pressure, blood volume, and extracellular fluids, for the transmission of nerve impulses and muscle contraction, for cellular exchanges and for the basic acid balance. Citric acid is a key molecule of the metabolic processes that take place within each cell of the body, and is present in important concentrations in bones, with a stabilizing function. In addition to

---

[33]Hans Adolf Krebs, a German-born British biologist (1900–1981) won the Nobel Prize in Medicine in 1953.

being produced by the body, it abounds mainly in citrus fruits and a little bit in all fruits, especially kiwis and strawberries.

Sodium citrate has applications both in the food industry, as an acidity correcting additive, and in the pharmaceutical industry, as an alkalizing compound against metabolic, gastric, and urinary acidosis; well-known, for example, is the use of sodium citrate in the prevention of urolithiasis from excess uric acid.

Synthesized through the fermentation of molasses by the fungus *Aspergillus niger*, in the food field sodium citrate is used as a corrector of acidity (buffer system) and flavor (slightly sour taste), antioxidant (prevents oxidation and subsequent browning of preserved fruit), chelator of metal ions, and nutrient for yeast in some fermented foods.

4.3.1.2 The mechanism of reduction

The use of citrate as a reducing agent for silver dates to the reaction of Turkevich used to obtain gold nanoparticles from chloroauric acid ($HAuCl_4$) [see Eq. (4.1)]. The Turkevich protocol is the most reliable and popular method for the synthesis of gold nanoparticles and consequently the recommended method by the National Institute of Standards and Technology for reference material preparation [15].

$$4Ag^+ + C_6H_5O_7Na_3 + 2H_2O \rightarrow 4Ag + C_6H_5O_7H_3 + 3Na^+ + H^+ + O_2 \ (4.4)$$

In Fig. 4.4 we can see the scheme of citrate reduction *in situ*.

In our method the synthesis has been modified with respect to the concentration. In order to compel the nanoparticles to grow on the fiber, we have tested various concentrations of $AgNO_3$ in order to find the correct concentration for a homogeneous film.

In short: to a stirred solution of $AgNO_3$ in water (2×10^{-3} M, 10 ml) at a temperature of 95°C, cotton yarn (10–15 mg) was added. After 10 minutes, trisodium citrate (3.42×10^{-2} M, 1 ml) was added portion wise. The clear colorless solution turned to pale yellow because of nanoparticles formation. After 30 min the functionalized yarn was taken out from the solution and rinsed repeatedly with water (5×20 ml) at room temperature. Eventual air drying at room temperature yielded dry fibers to be used without further manipulations.

In this case the problem is the high temperature. To act as a reducing agent the citrate must be heated to 95°C. In an industrial context this aspect would lead to a substantial change in the finishing plants. In addition, textile fibers, especially if already colored, can hardly stand these temperatures.

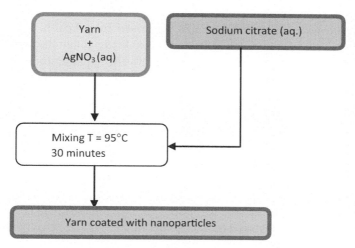

Figure 4.4   Scheme of citrate reduction in *in situ* synthesis.

For these reasons, the reduction with citrate has been discarded.

4.3.2   The Second One: Ascorbic Acid

4.3.2.1   Brief history

Until the 18th century scurvy was the nightmare of sailors: it caused bone swelling, fever, anemia, gingival bleeding, weight loss, muscle pain. Already in the 16th century, however, it was known that the cause was the lack of fresh vegetables and fruits. The sailors spent months and months on ships, far from the mainland, as happened to the soldiers, engaged on the front during the winter. No one, however, knew what made fresh fruits and vegetables the only possible remedy for scurvy.

The first experiments to understand which foods were able to prevent scurvy were conducted by James Lind (1716–1794), surgeon of the British Royal Navy. In 1747, inspired by the work of another British doctor, John Woodall (1570–1643), Lind subjected 12 sailors affected by scurvy to what is considered the first study-control study in history (a study where in addition to patients who experience a particular therapy there are others who are not undergoing treatment, and therefore act as a control group). To six of them Lind added a portion of oranges and limes to the daily ration and observed that this new diet improved their health. It took 6 years, however, before the surgeon was able to publish the results of his studies, and another 40 years before the Navy mandatorily added lemon or lime juice to the sailors' rations. Other foods, however, also seemed to work broccoli, cabbage, malt, and sauerkraut, imposed by James Cook (1728–1779) on sailors during his famous journey to Hawaii. It was clear, therefore, that there was an antiscorbutic agent (a term coined between the late 18th and early 19th centuries) in all these foods. In 1912, the Polish biochemist Kazimierz Funk (1884–1967) stated the concept of vitamins: "non-mineral nutrients essential for life to be taken with diet." The unknown agent was to be one of these vitamins and was baptized "vitamin C" in 1921.

In 1937 the Hungarian biochemist Albert Szent-Györgyi (1893–1986) won the Nobel Prize for the discovery of vitamin C.

Vitamin is the generic name of a heterogeneous group of organic substances belonging to bioregulators, indispensable, and irreplaceable for life, which humans and most animals take with food as such or in the form of precursors (provitamins) which are then activated by internal enzymatic factors or external factors (e.g., ultraviolet radiation of provitamin D).

The term coined by biochemist K. Funk (1884–1967) in 1912 to designate an aminic compound present in the outer layer of the caryopsis of rice and having the property of curing beriberi, was later extended to other substances not classifiable as plastic or energetic foods but indispensable, albeit in small quantities, for life, not always containing aminic groups. Vitamin C is an enantiomeric form of ascorbic acid, more precisely the levogyre form (see Fig. 4.5).

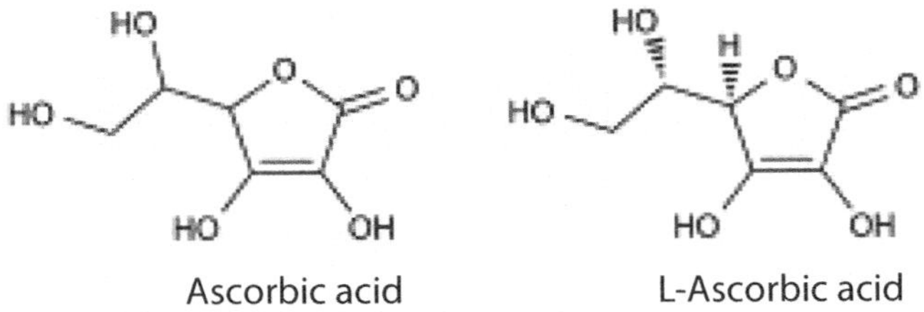

Ascorbic acid   L-Ascorbic acid

Figure 4.5 Ascorbic acid.

In 1934 Norman Haworth (1883–1950), an English biologist, successfully synthesized vitamin C in the laboratory, a work for which he won the Nobel Prize in Chemistry in 1937.

4.3.2.2 The mechanism of reduction

The mechanism of silver reduction by ascorbic acid is shown in Eq. (4.5).

$$Ag^+ + C_6H_9O_6 \rightarrow Ag + C_6H_8O_6 + H^+ \tag{4.5}$$

In our case the obtained AgNPs have a size around 50 nm, with a narrow size distribution [16].

The kinetics of reaction at room temperature in this case is lower than that of citrate, but the results is a very satisfying monolayer coating.

Also, in this case the concentration of $AgNO_3$ has been changed to obtain the best result. To a stirred solution of $AgNO_3$ in water (2 × 10^{-3} M, 10 ml) at a temperature of 40°C, cotton yarn (10–15 mg) was added. After 10 minutes, ascorbic acid (1 × 10^{-3} M, 10 ml) and trisodium citrate (3.42 × 10^{-2} M, 1 ml) were added portion wise (see Fig. 4.6).

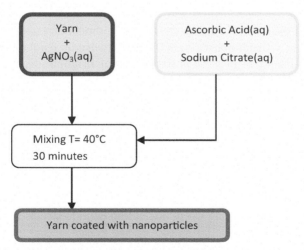

Figure 4.6   Scheme of ascorbic acid reduction in synthesis *in situ*.

The clear colorless solution turned to grayish because of nanoparticles formation. After 30 min the functionalized yarn was taken out from the solution and rinsed repeatedly with water

(5 × 20 ml) at room temperature. Eventual air drying at room temperature yielded dry fibers. Other cotton yarns were prepared in a similar fashion by using different ascorbic acid concentrations in the range 2×10^{-4} M to 1.43×10^{-2} M.

4.4 Experimental Section

At the state of the art, materials with antibacterial activity based on silver nanoparticles are prepared through distinct phases in which the preparation of nanoparticle suspensions (i) is followed by immersion of the fibers in these suspensions (ii) and subsequent post-treatment (iii). In this method, the fiber is simply the material to which the nanoparticles are to adhere. Moreover, these methodologies lead to a limited control of the level of silver nanoparticles coating, which is instead a desirable property, or in any case a lower control than that resulting from the application of our method.

In our original technique, fiber plays a fundamental and irreplaceable role in the process of initiation and growth of nanoparticles. It is specifically introduced into the reactor before nanoparticles are formed and used as a nucleation center donor. Without the fiber, the characteristics of homogeneity and continuity of coating with nanoparticles of the fiber itself would not be obtained.

The inventive process is based on our original idea to use the fiber itself as an initiator for the growth of nanostructured materials. This growth directly on the fiber leads to materials with special characteristics never seen before.

The simultaneous presence of the fiber and of the couple ascorbic acid/citrate anion is the necessary condition to carry out the synthesis of the silver nanoparticles possessing the desired characteristics.

The fiber has the function to promote the nanoparticles formation by providing many nucleation centers which are present on its surface.

The synergistic action of the above three factors leads to a regular nanoparticles distribution which is proved to be better than those obtained by dipping processes. For the same reason also the adhesion of the silver nanoparticles is significantly improved.

Ascorbic acid acts as the reducing agent by means of which the silver ions Ag^+ are transformed into metallic silver $Ag°$. Sodium citrate plays the fundamental role of stabilizer inhibiting silver nanoparticles collapse and allowing them to remain well separated from each other.

From an industrial point of view, the application of our process leads to the production of quality materials with defined characteristics without requiring substantial modifications to the production cycles commonly used in the textile industry. This simplicity of application translates into an economic benefit when you decide to produce these high value-added materials.

The applicability of our process to a production cycle in use has been demonstrated by tests on a pilot plant. The use of growth directly on fiber allows optimization of concentration, eliminating waste. Under appropriate conditions it is possible to minimize the presence of nanoparticles that do not nucleate and grow on the fiber.

In cases of dipping and soaking treatments with nanoparticles, it is extremely difficult for all the nanoparticles to adhere to the fiber. The remaining nanoparticles in solution are difficult to reuse. Not to be underestimated is also the lower amount of water required in our method, which combined with the saving of expensive reagents such as silver reduces production costs, including wastewater disposal costs.

4.4.1 Scanning Electron Microscope Images of Various Samples

In this section we report scanning electron microscope (SEM) images obtained from samples treated by varying reaction conditions. All images have been produced at the Nanolab laboratories of Energy Department of Politecnico di Milano.

In Fig. 4.7 are shown two images of different fibers on which silver nanoparticles have been nucleated and grown by using the couple ascorbic acid/sodium citrate according to our method. The fibers are uniformly and smoothly coated by separated nanoparticles with size around 50 nm.

The presence of the citrate as a stabilizer is extremely important. Although the citrate ion can act as a reducing agent in particular conditions, in our process—because it operates at room temperature—citrate is used as a stabilizer and not as a reducing agent.

Figure 4.7 Images obtained by a SEM of different cotton fibers containing silver nanoparticles manufactured according to our method.

In Fig. 4.8, it is shown the comparison between the images of two cotton fibers containing silver nanoparticles manufactured according to the method in absence of citrate, on the left, and in presence of citrate on the right panel. It is particularly evident how the presence of a stabilizer is essential to obtain a uniform distribution of nanoparticles avoiding the aggregation.

In the absence of citrate, the nanoparticles grow on the fiber resulting in a non-homogeneous coating formed by zones in which single nanoparticles are evident (type "B" zones) and agglomerates of particles of different sizes (type "A" zones). On the other hand, as stated before, in the presence of citrate the fibers are uniformly and smoothly coated by separated nanoparticles (that is the type "B" zone is not restricted).

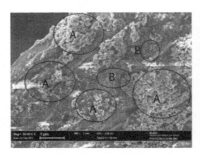 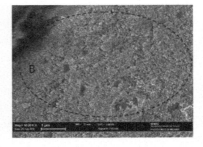

Figure 4.8 SEM image of a cotton fiber containing silver nanoparticles manufactured according to our method: (left) in absence of citrate and (right) in presence of citrate.

In the image of Fig. 4.9 is shown an example of the agglomerates of particles obtained after applying the process of the present application but operating in absence of the fibers and in absence of citrate. Such particles have mostly the size of hundreds of nanometers and are clustered in aggregates that have the size of micrometres. Many particles cannot be described as nanoparticles according to the characteristic of possessing at least one dimension between 1 and 100 nm. In absence of the fibers that provide many nucleation centers there is no nanoparticles formation.

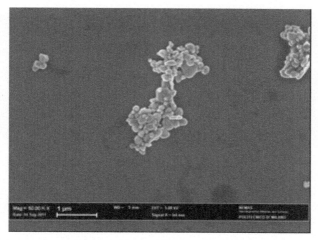

Figure 4.9 SEM image of a cotton fiber containing silver nanoparticles manufactured according to the method in absence of citrate.

The role of the couple ascorbic acid/sodium citrate on the morphology of the nanoparticles has been studied in our laboratory. In particular, we have investigated thoroughly the influence of the ratio of their concentrations in order to find the most favorable process conditions since these concentrations are very critical.

After fixing a silver salt solution concentration we have applied the process by using variable concentrations of ascorbic acid and of sodium citrate thus varying the concentration ratio of ascorbic acid to sodium citrate.

In the below reported example it is demonstrated the remarkable influence of the concentration ratios of ascorbic acid to sodium citrate on the manufacturing of the fibers containing silver nanoparticles.

The selected starting concentrations in this set of experiments are 2 mM for silver salt and 34.2 mM for sodium citrate.

In Fig. 4.10 are shown two images of a fiber containing silver nanoparticles manufactured by the process in which the concentration ratio of ascorbic acid to sodium citrate are, respectively, 5.85×10^{-3} M on the left and 2.92×10^{-2} M on the right. It is evident that the manufactured materials do not meet the expected requirements in terms of homogeneity and size dispersion of the silver nanoparticles.

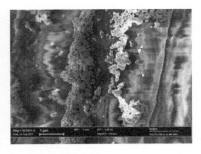

Figure 4.10 SEM images of a fiber containing silver nanoparticles manufactured according to the method of the present application (starting concentrations: silver salt = 2 mM, sodium citrate = 34.2 mM) with different ascorbic acid to sodium citrate concentration ratio: 5.85×10^{-3} M on the left and 2.92×10^{-2} M on the right.

In Fig. 4.11 are shown two images of a fiber containing silver nanoparticles manufactured by the process of the present application in which the concentration ratio of ascorbic acid to sodium citrate are, respectively, 4.15×10^{-3} M on the left and 7.07×10^{-2} M on the right. The manufactured materials show an acceptable size dispersion and distribution of the nanoparticles, however, they do not meet the expected requirements in terms of covering.

In Fig. 4.12 are shown two images of a fiber containing silver nanoparticles manufactured by the process of the present application in which the concentration ratio of ascorbic acid to sodium citrate are, respectively, 1.46×10^{-1} (upper panel on the left), 2.60×10^{-1} (upper panel on the right) and 4.18×10^{-1} (lower panel). All the manufactured materials show an excellent carpet-like situation that meet the expected requirements in terms of size dispersion, distribution and separation of the nanoparticles, and covering of the fiber.

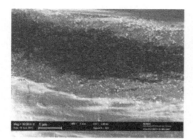 

Figure 4.11 SEM images of a fiber containing silver nanoparticles manufactured according to the method of the present application (starting concentrations: silver salt = 2 mM, sodium citrate = 34.2 mM) with different ascorbic acid to sodium citrate concentration ratio: (left) 4.15×10^{-2} M and (right) 7.07×10^{-2} M.

It is remarkable that such a kind of nanoparticle size which is normally obtained with strong reducing agents (e.g., sodium borohydride) has been obtained with a weak reducing agent such as ascorbic acid. This is an evidence of the major role played by the fiber in promoting the nanoparticles formation by providing many nucleation centers which are present on its surface.

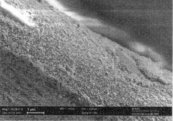

Figure 4.12 SEM images of a fiber containing silver nanoparticles manufactured according to the method of the present application (starting concentrations: silver salt = 2 mM, sodium citrate = 34.2 mM) with different ascorbic acid to sodium citrate concentration ratio: 1.46×10^{-1} (upper left), 2.60×10^{-1} (upper right), and 4.18×10^{-1} (lower).

In summary to highlight the importance and innovation of the WO2009132798A1 patent, a morphological comparison is made using SEM between nanoparticles in solution and nanoparticles grown on cotton fiber.

Nine samples were prepared with different concentration ratios of stabilizer (sodium citrate) and reducing agent (ascorbic acid). The reaction temperature was kept constant at 30°C, magnetic stirring was maintained throughout the reaction period, and the reagent introduction times were the same for all the prepared samples.

The different concentration ratios are shown in the Table 4.1.

Table 4.1 Scheme of different concentration of reducing agent, and stabilizer with respect to the concentration of silver salt kept constant

Sample	$[AgNO_3]$	[Citrate]	[Asc. Acid]	[Asc. Acid]/ $[AgNO_3]$	[Asc. Acid]/ [citrate]
2	2×10^{-3} M 10 ml	-	1.42×10^{-3} M 10 ml	0.71	1.42×10^{-3}
3	2×10^{-3} M 10 ml	3.42×10^{-2} M 1 ml	1.42×10^{-3} M 10 ml	0.71	4.15×10^{-2}
4	2×10^{-3} M 10 ml	-	2.42×10^{-3} M 10 ml	1.21	2.42×10^{-3}
5	2×10^{-3} M 10 ml	3.42×10^{-2} M 1 ml	2.42×10^{-3} M 10 ml	1.21	7.07×10^{-2}
8	2×10^{-3} M 10 ml	3.42×10^{-2} M 1 ml	1.00×10^{-3} M 10 ml	0.50	2.92×10^{-2}
9	2×10^{-3} M 10 ml	3.42×10^{-2} M 1 ml	2.00×10^{-4} M 10 ml	0.10	5.85×10^{-3}

[Ascorbic Acid]/[Ag] ratio determines nanoparticles dimension and density on the fiber. This parameter has been related, among the others, to AgNPs nucleation ratio in colloidal synthesis [17]. Moreover, high values of [Ascorbic Acid]/[Ag] ratio influence the formation of Ag ascorbate, this complex acts as a seed further increasing nanoparticles nucleation on the fiber [18].

We have investigated thoroughly the influence of this ratio, in order to find the most favorable process conditions, ranging it between 0.1 and 7.1.

In Fig. 4.13a is shown a fiber, produced using our method, in which the [Ascorbic Acid]/[Ag] ratio ranges between 0.1 and 0.5. AgNPs are aggregated in a non-homogeneous way, moreover their dimension is, as in the case of citrate absence, hundreds of nanometers.

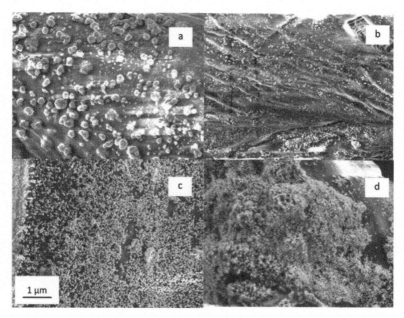

Figure 4.13 SEM images of cotton fibers treated with varied concentration ratio [Ascorbic Acid]/[Ag] (a) 0.1–0.5 range, (b) 0.7–1.2 range, (c) 2.5–4.5 range, and (d) 7.1 ratio.

Raising value of the [Ascorbic Acid]/[Ag] ratio over the stoichiometric amount (0.5) in the (0.7–1.2 range) induces the nucleation of small particles (25–30 nm), see Fig. 4.13b the manufactured material shows an acceptable size dispersion and distribution of AgNPs, however, nanoparticle coverage its very poor.

In Fig. 4.13c it is shown a fiber decorated by Ag nanoparticles in which the [Ascorbic Acid]/[Ag] ratio is (2.5–4.5). Nanoparticles covering the fiber are uniformly distributed, very dense with a narrow size distribution of dimension.

When the [Ascorbic Acid]/[Ag] ratio is high, 7.1, the dimension of nanoparticles is still around 30 nm but huge non-uniformities in the fiber coverage take place (Fig. 4.13d).

A similar behavior has been seen in the production of Ag nanoparticles (300 nm) on a polymer inclusion membrane, were the loaded Ag ions have been reduced by ascorbic acid. Raising the [Ascorbic Acid]/[Ag] ratio results in an increasing of the number of nanoparticles until the aggregation in large clusters takes place [19].

We can see in detail (see Fig. 4.14) how much the stabilizer influences the nucleation and growth of silver nanoparticles. [citrate]/[Ag] ratio is related to particle dimensions and shape [20]. In Fig. 4.14, in absence of citrate, in these conditions the nanoparticles grow on the fiber resulting in a non-homogeneous coating formed by a mixture of single nanoparticles and agglomerates of different size (hundreds of nm), the presence of the citrate, as a stabilizer, is extremely important. According with literature we choose [citrate]/[Ag] ratio at a value of 1.7 in order to ensure small dimensions and round shape.

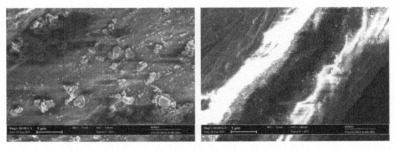

Figure 4.14   Influence of stabilizer in the solution: (left) without citrate and (right) with citrate.

4.4.1.1   Difference between dipping and *in situ*

We have seen in Chapter 3 the treatments that allow the covering of textile materials with silver nanoparticles. We are essentially talking about dipping, reduction by UV rays from silver salt solutions and coating by physical methods.

In our method, the material to be coated becomes nucleation center for the nanoparticles. In this way, unlike a dipping in which silver nanoparticles attach themselves to the material by electrostatic forces, in our method the nanoparticles grow directly on the fiber, maintaining a well-defined shape and above all allowing an almost monolayer coating. The use of weak reducing agents such as ascorbic acid and citrate would result in a slower reduction with a higher growth of nanoparticles. The presence of a fiber or material in solution drastically accelerates the reduction of Ag^+ during the growth stage.

In Figure 4.15 we can see the scheme of a standard procedure to cover a fiber by dipping.

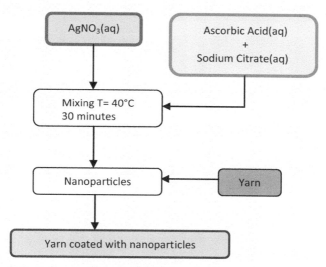

Figure 4.15   Scheme of dipping procedure.

In order to prove the different growth of nanoparticles on the fiber in the patented synthesis compared to a coating of the fiber by dipping in a colloidal solution of silver nanoparticles, we immersed the fiber in a colloidal solution prepared by reducing a silver salt with ascorbic acid using the same operating conditions as fiber synthesis.

The SEM images of the fiber show a very different coating compared to growth *in situ*. With the dipping procedure, the nanoparticles aggregate and form complex structures (Figure 4.16). On the contrary, using the claimed process, the coating is uniform, and the nanoparticles do not aggregate (Figure 4.17). This is very important if one considers that the smaller the particles—and thus the higher the surface/volume ratio—the higher their antibacterial activity [21].

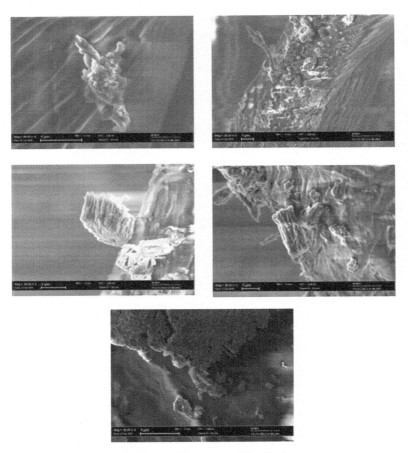

Figure 4.16 Images of polyester fibers treated by dipping.

Briefly the procedure followed for dipping ascorbic reduction synthesis (dipp-ARS) to a stirred solution of $AgNO_3$ (2×10^{-3} M, 10 ml) at a temperature of 40°C, ascorbic acid (1.42×10^{-3} M, 10 ml) and TSC (3.42×10^{-2} M, 1 ml) were added portion-wise. After 5 minutes, the fibers (10–15 mg) were dipped into the above grayish suspension and the system was let to cool to room temperature. After 24 hours the functionalized yarn was taken out from the solution and rinsed repeatedly with water (5×20 ml) at room temperature. Eventual air drying at room temperature yielded dry fibers.

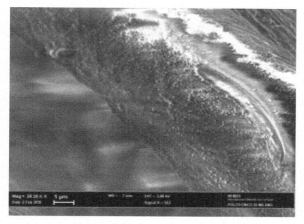

Figure 4.17   Cotton yarn treated with silver colloidal growth.

4.5   Different Ways of Application: Wet and Spray

After submitting the patent application, the difficult part was to industrialize the process.

The industrial part of the request was to modify the plant already in use as little as possible.

The best solution seemed to be to follow the methods of applying the finish. Through the application of different types of products, it is possible to give a fabric certain property that would be impossible to achieve using only mechanical processes. It is thus possible to make fabrics with temporary mechanical finishes, such as calendaring, stable, and achieve, among the many possible, waterproofing, or fireproofing properties on fabrics otherwise lacking in them. The application methods of these substances are different: solution, dispersion, emulsion by foulard impregnation, exhaustion, coating, spraying, and other methods.

The application of the finishing substances on the fibers is for the most part conditioned by the hygroscopic and structural characteristics of the textile material, the desired effect, the physical and chemical nature of the components of the finishing substances and the production rate. The following main types of application, recurring in textile finishing, can be differentiated:

(a) procedure for scarfing.
(b) spraying procedure with spraying device.
(c) procedure for long bath exhaustion.
(d) spraying procedure with doctor blades.
(e) controlled application of low liquor quantities.

We started with the application in solution. The application in solution is very convenient in the laboratory but from an industrial point of view there may be some problems in the development of the method.

Due to a potential customer's request, we switched to spray application. Let us see the two procedures in more detail.

4.5.1   By Solution

The application in solution can be done in two ways both in dyeing tubs and in dyeing cabinets. In the first case we should provide at least two tanks containing the two types of solution. The material to be treated can be passed on conveyor belts.

The negative aspect is that the solution of the reducing agent will have to be changed with a certain frequency, the material soaked in silver salt will drip a little in the tank decreasing the concentration of ascorbic acid available.

The second way that has been experimented is the positioning of a wire spool in an autoclave (see Fig. 4.18 on the left). Initially the silver solution is loaded and passed through the strand, as is done for any finishing. The next step involves the unloading of the solution and the subsequent loading of the silver solution, which is also fluxed through the fortress. The result visible in Fig. 4.18 on the right is more than satisfactory, the difference between the wire spool treated and not treated is clear.

The dyeing cabinet (see Fig. 4.19) is a device used for dyeing hanks, consisting of a parallelepiped tub divided into compartments by perpendicular partitions. The hanks are arranged on special material holders, which can be hooked inside the machine in special grooves; the bath circulates in both directions (ascending and descending) and the material, not being packed, does not offer great resistance.

Figure 4.18   Example of autoclave (left) and wire spool (right).

Figure 4.19   Dyeing cabinet.

To reduce the amount of solution to be used, a particular device has been created, using the utility model [22], which allows you to reduce the amount of liquid to be used in these cabinets (Fig. 4.20).

In short, we have three containers (A, B1, B2). The first (A) is inside the dyeing cabinet (AT) and is loaded with textile fibers (FIB), in the form of spools and/or hanks (M), to be functionalized. The second (B1) is filled with a aqueous solution of silver nitrate and the third (B2) is filled with an aqueous solution of citrate and ascorbic acid. The containers are removable and allow to reduce the amount of solution.

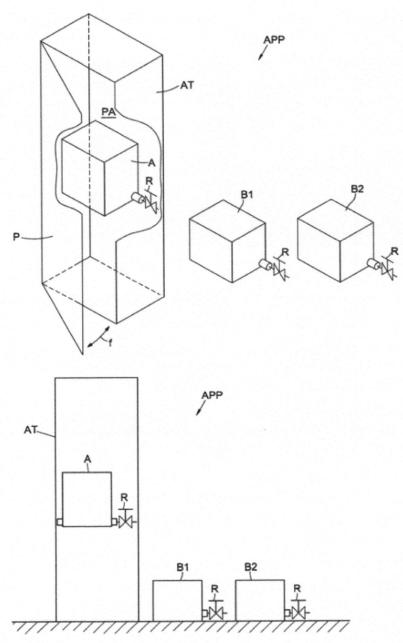

Figure 4.20 Scheme of our utility model.

The equipment allows advantageous recovery of $AgNO_3$ silver nitrate and ascorbic acid plus sodium citrate solutions that have not been absorbed by the textile fibers and is also suitable to conveniently implement, using already known equipment such as a usual textile dyeing cabinet, an industrial process aimed at producing textile fibers functionalized with silver nanoparticles and therefore exhibiting deodorizing and bactericide properties.

4.5.2 By Spraying

Over the years, a mattress cover manufacturer has been interested in licensing our patent. The plant was completely different from those of a dyeworks. First, it was fabric and not yarn, the so-called non-woven fabric, so the material would run on rollers and it would not be possible to wet it because it would become too heavy. Then the company in question wanted to apply the method in a way that would not have too much impact on their plants.

Our solution was to apply two sets of sprayers so that the fabric would continue its run without being obstructed (see Fig. 4.21). Two 25-liter canisters were loaded: one was filled with silver nitrate solution, the other with ascorbic acid and citrate solution. It was important to use PP tanks and not common galvanized tanks in order to avoid the reduction of silver by zinc.

Figure 4.21 (Left) The plant before, and (right) the plant with the double series of sprayers.

The first row of nozzles sprays a solution of silver nitrate and sodium citrate in water, the second row sprays an aqueous solution of ascorbic acid.

The fabric (polypropylene) flows horizontally to the sprayed side at a speed of 20 m/min, then it is passed under a series of IR lamps (temperature about 190°C) and comes out almost completely dry (see Fig. 4.22).

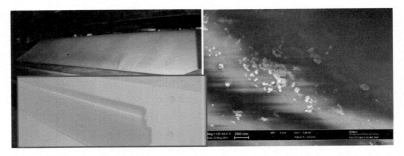

Figure 4.22 After antibacterial treatment on the right SEM images of the fabric.

Antibacterial tests have shown good results, demonstrating the effectiveness of the treatment by spraying.

Last but not least, an application that is a middle ground. A label and packaging company needed a solution for treating both the tissue in which shirts are wrapped and the labels to be applied to the garments. Once again, it was not possible to change the type of plant, the solution was by no means a conclusion not obvious.

Initially, tests were carried out in the laboratory, mixing the ink with our solutions (see Fig. 4.23).

Figure 4.23 Examples of (left) mixed solutions, (middle) the paper, and (right) the printed paper.

The material is printed at a printing plant that is normally used by the company, combining the normal process with our modified methodology. No additional steps or tools are included (see the picture of printing plant used in Fig. 4.24).

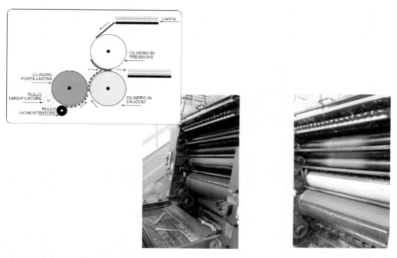

Figure 4.24 Offset printing machine, on the right with the printed color.

To verify the presence of silver nanoparticles in the printed ink, we have printed ink after the test on pilot plant (see Fig. 4.25).

Figure 4.25 (Left) Ink as is on paper. (Right) Printed with ink containing silver, increasing silver concentration from 1 to 5.

This time we used a special technique to analyse the material: atomic force microscopy (AFM) [23], widely used to evaluate surface roughness.

In Figure 4.26 we can see the AgNPs presence also on the printed paper.

It is not easy to understand AFM images, there are not as clear as those obtained by SEM. In our case the difference between treated and untreated paper is quite visible.

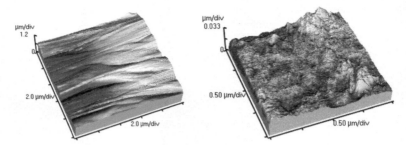

Figure 4.26 AFM images of printed paper: (left) only paper and (right) printed paper with AgNPs.

References

1. Lahann, J., Mitragotri, S., Tran, et al. *Science* (2003) **299**, p. 371.

2. Muhammad Raffi, M., Mehrwan, S., Bhatti, T.M., et al. *Annals of Microbiology* (2010) **60**, pp. 75–80.

3. Visai, L., De Nardo, L., Punta, C., et al. *International Journal of Artificial Organs* (2011) **34** (9), pp. 929–946.

4. Sirelkhatim, A., Mahmud, S., Seeni, A., et al. *Nano-Micro Letters* (2015) **7** (3), pp. 219–242.

5. Lee, K. J., Browning, L. M., Nallathamby, P. D., and Xu, X.-H., *Chemical Research in Toxicology* (2013) **26** (6), pp. 904–917.

6. Zandi-Atashbar, N., Hemmateenejad, B., and M. Akhond, *Analyst* (2011) **136** (8), pp. 1760–1766.

7. Mulfinger, L., Solomon, S. D., Bahadory, M., et al. *Journal of Chemical Education* (2007) **84** (2), pp. 322–325.

8. Abbaszadegan, A., Ghahramani, Y. et al. *Journal of Nanomaterials* (2015), p. 720654.

9. Samia Al Azharia Jahn *Anthropos* (1999) **94**, pp. 419–430.

10. Bannea, S.V., Patilb, M.S., Kulkarnic, R.M., and Patilb, S.J. *Materials Today: Proceedings* (2017) **4**, pp. 12054–12060.

11. Nickel, U., Castell, A., Pöppl, K., and Schneider, S. *Langmuir* (2000) **16**, pp. 9087–9091.

12. Liz-Marzan, L.M., and Lado-Tourin, I. *Langmuir* (1996) **12**, pp. 3585–3589.

13. Panigrahi, S., Kundu,S., Ghosh, S., Nath, S., and Pal, T. *Colloids and Surfaces A: Physicochem. Eng. Aspects* (2005) **264**, pp. 133–138.

14. (a) Krebs, H. A. and Johnson, W. A. *Enzymologia* (1937) **4**, pp. 148–156. (b) Nobel Lecture, December 11, 1953.

15. Turkevich, J., Cooper Stevenson, P., and Hillier, J. *Discussion Faraday Society* (1951) **11**, pp. 55–75.

16. Kimling, J., Maier, M., Okenve, B., Kotaidis, V., Ballot, H., and Plech A., *Journal of Physical Chemistry B* (2006) **110**, pp. 15700–15707.

17. Suber, L., Sondi, I., Matijevic, E., and Goia, D.V. *Journal of Colloidal Interfaces Science* (2005) **288**, pp. 489–495.

18. Fukuyo, T., and Imai, I. *Journal of Crystal Growth* (2002) **241**, pp. 193–199.

19. Bonggotgetsakul, Y.Y.N., Cattrall, R.W., and Kolev, S.D. *Journal of Membrane Science* (2013) **428**, pp. 142–149.

20. Henglein, A., and Giergig, M. *J. Phys. Chem. B* (1999) **103**, 9533–9539.

21. (a) Baker C., Pradhan A., Pakstis I., et al. *Journal of Nanoscience Nanotechnology* (2005) **5**, pp. 244–249. (b) Panácek, A., Kvìtek, I., Prucek R., et al. *Journal of Physical Chemistry B* (2006) **110**, pp. 16248–16253.

22. Facibeni, A., Bottani, C.E., Dellasega, D., and Giribone, D. Equipment, dyeing cabinet type for the functionalisation of textile fibre with silver nanoparticles, ITUB201543336U1 (2015).

23. Voigtländer, B. *Atomic Force Microscopy* (2019), Springer.

Chapter 5

Antibacterial Properties of Silver Nanoparticles

5.1 Introduction

We have already said that silver nanoparticles, AgNPs, have been used since ancient times as powerful antibacterial agents. But how did the ancients who did not have scientific tests to evaluate it notice it? They tested them directly on the diseased, probably looking at the results.

There are traces that ancient civilizations used a variety of naturally available treatments for infections, mainly herbs, honey, and in some cases even animal feces [1].

Incredibly, some more modern antibiotics have also been found. Traces of tetracyclines have been found in human skeletons excavated in Nubia and during the Roman occupation of Egypt [2]. The origin of the tetracycline remains a mystery.

The existence of a world inhabited by extremely small organisms had been intuited since ancient times. The Latin writer Marco Terenzio Varrone (116 BC–27 BC) spoke of "animalia quaedam minuta quae non possunt oculis consequi"[34] capable of causing marsh fever. Lucretius (98 BC–55 BC) in *De rerum natura* speculated

---

[34]animals so small that they cannot be seen with the eyes.

Silver Nanoparticles: Synthesis, Properties, and Applications
Anna Facibeni
Copyright © 2023 Jenny Stanford Publishing Pte. Ltd.
ISBN 978-981-4968-21-8 (Hardcover), 978-1-003-27895-5 (eBook)
www.jennystanford.com

philosophically about the existence in the air of "seeds" harmful to man.

These ideas were taken up and developed after more than 14 centuries by the Italian physician G. Fracastoro (1476–1553), who for the first time enunciated the concept of the transmission of infections in humans and animals by invisible living organisms. The first concrete evidence of the existence of the microbial world was provided by A. van Leeuwenhoek (17th–18th century), a cloth trader from Delft (Holland), with a simple microscope he built himself. He observed and described in 1676 beings infinitely small and varied in shape and size which he referred to as *animalcula*. However, the progress of knowledge about very small living beings was hindered by beliefs and superstitions rooted in antiquity concerning their possible spontaneous generation.

In the late 1800s, Robert Koch[35] (1843–1910) and Louis Pasteur[36] (1822–1895) were able to establish the association between individual species of bacteria and disease through propagation on artificial media and in animals. Alexander Fleming (1881–1955) discovered penicillin in 1928 [3].

Initially, the best source of new agents was from other naturally occurring microorganisms and after Streptomycin [4] was isolated in 1944 from *Streptomyces griseus* (an organism found in soil), a worldwide search began.

From the mycetes *Cephalosporium acremonium* the Italian doctor Giuseppe Brotzu (1895–1976) isolated Cephalosporin C. Cephalosporins are widely used therapeutically against infections sustained mainly by gram-positive bacteria, but also by gram-negative bacteria. Cephalosporins are generally divided into "generations," which should not be understood in temporal terms but should be referred to the spectrum of action. Cephalosporins act, in part, on the same protein targets as penicillin. They are able to inhibit bacterial growth by interfering with bacterial wall [5].

---

[35]discovered the bacteria that caused anthrax, septicaemia, tuberculosis, and cholera, and his methods enabled others to identify many more important pathogens. His first important discovery was on anthrax, a disease that killed large number of livestock and some humans.

[36]is best known for inventing the process that bears his name, pasteurization. Pasteurization kills microbes and prevents spoilage in beer, milk, and other goods.

To evaluate the *in vitro*[37] antimicrobial activity of an extract or a pure compound we can use a variety of laboratory methods. The most known and basic methods are the disk diffusion and broth or agar dilution methods.

Other methods are used especially for antifungal testing, such as poisoned food technique and TLC-bioautography. To further study the antimicrobial effect of an agent in depth, time-kill test, and flow cytofluorometric methods are recommended, which provide information on the nature of the inhibitory effect (bactericidal or bacteriostatic), if this is time-dependent or concentration-dependent, and the cell damage inflicted to the test microorganism.

In vitro testing occurs in a laboratory and usually involves studying microorganisms or human or animal cells in culture. This methodology allows scientists to evaluate various biological phenomena in specific cells without the distractions and potential confounding variables present in whole organisms.

In vivo testing, especially in clinical trials, is a vital aspect of medical research in general. *In vivo* studies provide valuable information regarding the effects of a particular substance or disease progression in a whole, living organism. The main types of *in vivo* tests are animal studies and clinical trials.

Antimicrobial susceptibility testing (AST) is a laboratory procedure performed by medical technologists (clinical laboratory scientists) to identify which antimicrobial regimen is specifically effective for individual patients. These are required by the International Organisation for Standardisation (ISO) [6].

5.2 Kind of Bacteria

Bacteria are monocellular microorganisms belonging to the domain of prokaryotes. The bacteria have dimensions that can vary from 0.2 to 30 µm [7].

Bacteria can be classified according to their shape: cocci (spherical), bacilli (stick), spirilli (spiraliform), vibrions (comma-shaped) [8].

We can classify them according to their growth temperature:

Thermophilic bacteria: grow at high temperatures (47–70°C, optimum temp. 50–55°C)

---

[37]*in vitro* is Latin for "in glass," *in vivo* is Latin "within the living"

Mesophilic bacteria: grow at intermediate temperatures (20–45°C, optimum temp. 30–37°C)

Psychrophilic bacteria: grow at low temperatures (0–25°C, optimum temp. 20–25°C).

We can classify them according to their respiration. Aerobic bacteria if they use oxygen and Anaerobic bacteria if they do not use oxygen.

They can be also pathogens, symbionts, commensals, and opportunists.

Another extremely important classification method is based on the reaction to Gram staining, a laboratory test developed by a Danish doctor, H. J. C. Gram (1853–1938); according to this criterion gram-positive (also gram+) and gram-negative (gram–) bacteria are distinguished [9].

Gram staining is a procedure used to detect the presence of bacteria and sometimes fungi in samples from sites of suspected infection. This test provides results quickly and, in addition to detecting the presence of microorganisms, also provides an initial classification.

The technique involves differential staining (two dyes are used during the procedure) after which, under the light microscope, gram positives appear violet in color while gram negatives are pink. The procedure involves four basic steps:

First step: treat the test sample (previously fixed on a slide) with crystal violet[38] dye for 3–5 minutes.

Second step: remove the dye by washing the test sample with iodine solution and let it act for about 2 minutes.

Third step: decolorize the slide with acetone for 5 seconds, taking care to rinse it immediately with water.

Fourth step: treat the slide with the second contrasting dye (safranin[39]) and let it act for 30 seconds, after which wash the slide with water and let it air dry.

By light microscope, it will be observed that bacteria that take the color of the second dye are gram-negative, while the others,

---
[38]A basic dye also known as gentian violet
[39]A cationic dye

stained with crystal violet, are gram-positive. What is it that allows gram-positives and -negatives to take on a different coloration? The answer lies in the different chemical composition of the bacterial wall.

Fungi (in the form of yeast or mold) can also be recognized by Gram stain, but the same is not true for viruses.

Gram staining is therefore a useful first level examination aimed at the detection and generic identification of the type of bacteria or fungi present. The results, although preliminary, are useful in the choice of tests to be performed and sometimes in the therapies that can be administered.

5.3   Antibiotics

The name antibiotic comes from the Greek ἀντί, against and βιος life. It was defined as a natural substance produced by a microorganism that was able to inhibit the growth of other microorganisms. This definition has now been replaced by a substance produced by a microorganism or a similar substance (produced wholly or by chemical synthesis) which at low concentrations inhibits the growth of other microorganisms.

Antibiotics (such as penicillin) are those produced naturally (by one microorganism fighting another), whereas nonantibiotic antibacterial (such as sulphonamides and antiseptics) are fully synthetic.

An antibiotic should be selective, bactericidal, have a broad spectrum of activity, be nontoxic to the host, have a long half-life in plasma, good tissue distribution, low plasma protein binding, oral and parenteral administration, no interference with other drugs, selective toxicity, and no local and/or systemic side effects.

When we talk about bacterial resistance, we mean the phenomenon that bacteria can survive and multiply in the presence of an antibacterial drug. From a clinical point of view, we speak of chemoresistance when pathogenic bacteria are not inhibited by an antibacterial drug at the site of infection and the drug is therefore ineffective.

Antibiotics can be classified according to the type of activity they exert on bacteria:

> ➤ Antibiotics with bactericidal activity: capable of killing the bacterial cell (MCB).
> ➤ Antibiotics with bacteriostatic activity: capable of inhibiting bacterial cell multiplication (MIC).

Let us see what MIC and MCB mean.

MIC (minimum inhibiting concentration): It is the lowest concentration of the test compound required to inhibit the growth of a given organism.

MBC (minimum bactericidal concentration): It is the lowest concentration of the test compound required to cause the death of more than 99.9% of a given organism.

The bacteriostatic or bactericidal action of an antibiotic depends on the mechanism of action. Antibiotics that act on structures fundamental to the bacterial cell such as the wall or nucleic acids will be bactericidal. If the antibiotic is bactericidal the MIC and MBC values are the same. If the antibiotic is bacteriostatic the MIC and MBC values are different (MBC > MIC).

5.3.1   How to Evaluate the Efficacy of an Antibiotic?

Resistance to antibiotics is a natural defense mechanism of bacteria. In recent years, the phenomenon of antibiotic microbial resistance (AMR) has increased dramatically, with each microorganism causing diseases of varying severity and incidence and against which there may be few or many effective chemotherapies or even other forms of primary prevention such as vaccination.

The problem of AMR is complex as it has several causes: the increased use of these drugs (including inappropriate use) in both human and veterinary medicine; the use of antibiotics in animal husbandry and agriculture; the spread of hospital infections caused by antibiotic-resistant microorganisms (and the limited control of these infections); and the increased spread of resistant strains due to increased international travel and migration flows.

The continued use of antibiotics increases the multiplication and spread of resistant strains. In addition, the emergence of pathogens resistant to several antibiotics at the same time (multidrug resistance) further reduces the possibility of effective treatment. It should be emphasised that this phenomenon often concerns healthcare-related infections, which arise and spread within hospitals and other healthcare facilities.

As time moved along, resistant gram-positive infections such as MRSA and enterococci were proving increasingly more challenging to clinicians, so antibiotic development shifted attention toward these bacteria.

WHO, at the World Health Assembly (2015), adopted the Global Plan of Action (GAP) to address antimicrobial resistance by setting five strategic goals aimed at:

- improve awareness levels through effective information and education targeting healthcare workers and the general population
- strengthen surveillance activities
- improve infection prevention and control
- optimize the use of antimicrobials in the field of human and animal health
- support research and innovation.

The European Union, which has been committed to combating antibiotic resistance for many years, developed its new Action Plan to Combat Antibiotic Resistance in 2017, based on a "One Health" approach that considers human, animal, and environmental health in an integrated manner.

In clinical practice, antibiogram is a report essential for choosing the correct therapy against infections. It makes possible to see which drugs are most effective against a certain pathogenic microorganism. Evaluation of the *in vitro* sensitivity of an antibiotic to a pathogenic germ also provides an estimate of the most appropriate therapeutic dose for treating the infectious disease.

You will therefore understand how important it is to detect the antimicrobial activity of an agent. It was therefore important to find a procedure to standardize the test so as not to obtain conflicting results.

The various techniques for performing this type of test can basically be traced back to two main methods: the agar diffusion method and the progressive dilution method, which we will look at in more detail in the following pages.

Another test used to check the effectiveness of an antibiotic is one that combines an analysis and purification technique such as thin-layer chromatography with a biological test. We are talking about TLC bioautography.

Let us see some of these techniques in detail.

5.4 Diffusion Methods

The method of diffusion on a plate provides that the microorganism is suspended in sterile distilled water until it reaches a certain turbidity and then sown by surface plugging on plates of agar medium on which are then placed nitrocellulose discs soaked in the substance to be tested. The culture media are then maintained at 35–37°C for approximately 18–24 hours.

After 24 hours, the halo that has formed around the antibiotic diskette is evaluated. The diameter of the halo in millimetres is related to the sensitivity of the bacterium to the antibiotic. The greater the halo, the greater the sensitivity of the bacteria to the antibiotic. The smaller the halo, the more resistant the bacterium is to the antibiotic.

This is a quali-quantitative method commonly used in the laboratory for the rapid evaluation of the efficacy of different substances against a microorganism.

Currently, the most widely used disk diffusion test is the Kirby–Bauer method, developed in the early 60s [10].

It is the most commonly used procedure in the laboratory and allows to obtain an assessment of MIC, which is the minimum concentration of antibiotic capable of inhibiting bacterial growth.

This method is also quantitative, so it allows to accurately determine in addition to the MIC also the MBC, which is the lowest concentration of antibiotic capable of destroying all bacteria. The method is valid and precise, but unfortunately also expensive and long implementation, so the use is limited to a few cases.

This method is particularly suitable when treating fast-growing aerobic microorganisms.

The powdered culture medium consists of meat extract, casein acid digest, starch, and agar in appropriate proportions.

5.4.1 Antimicrobial Gradient Method (E-test)

The antimicrobial gradient method (also known as epsilometer test or E-test) is a technique which is able to combine dilution and diffusion methods. It is possible to establish the MIC by creating a concentration gradient of the antimicrobial agent tested in the agar medium.

The E-test utilizes rectangular plastic strips treated with a well-defined, continuous, and increasing gradient of antibiotic concentration, with the maximum value at the end of the strip. A two-letter code specifies the type of antibiotic.

Application of the E-test[40] strips to solid medium (a Mueller–Hinton agar inoculated homogeneously and in a standardized manner with the bacterial strain to be tested) results in the formation of a gradient of the drug in the surrounding medium.

During the 1950s, the scientific founder of AB BIODISK Hans Ericsson (professor of microbiology at the Karolinska Hospital and Karolinska Institute, Stockholm) developed this method to standardize the disc diffusion method and to enhance its reproducibility and reliability.

The E-test strip was first described in 1988 and was introduced commercially in 1991 by AB BIODISK. The French company BioMérieux acquired AB BIODISK in 2008 and continues to manufacture and market this product range under the mark E-test. BioMérieux operates in the medical sector—microbiology and diagnostics—and a world specialist in *in vitro* diagnostics.

E-test was first presented at the Interscience Conference on Antimicrobial Agents and Chemotherapy (ICAAC) in Los Angeles in 1988 as a novel gradient system for MIC determinations. In September 1991, E-test was launched globally as a MIC product after receiving the USA Food and Drug Administration (FDA) clearance.

It is an using plastic strips impregnated with a concentration gradient (predefined exponential gradient) of antibacterial drug (Fig. 5.1).

After 48 hours incubation the MIC is read as the concentration (mg/ml) of antibacterial drug indicated by the strip at the point of intersection with the elliptical bacterial growth inhibition halo.

---

[40]https://www.biomerieux.it/prodotto/etestr

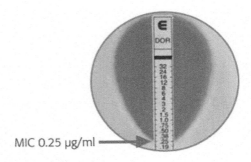

MIC 0.25 µg/ml

Figure 5.1   E-test by BioMérieux.

5.5   Dilution Methods

Are generally better methods for accuracy, standardization, and reproducibility; when performed with a large number of dilutions and numerous microbial strains they are in most cases used for the evaluation of new antimicrobial agents.

In this technique, microbial growth is assessed in culture media to which scalar concentrations of the antimicrobial agent have been added; it is important to avoid using culture media that may inhibit the activity of some antimicrobial agents. The antimicrobial agent used should be sterile, properly stored, and of known activity.

With dilution methods it is possible to evaluate the combination of two or more drugs. The advantages of dilution methods are the possibility of accurately calculating the minimum inhibitory dose, and in the case of liquid media, the microbicidal activity.

The culture medium can be liquid or solid.

In liquid culture media, also called culture broths, nutrients are dissolved in water. The growth of bacteria in this type of medium can be demonstrated by the appearance of turbidity.

Solid culture media are obtained by adding a gelling agent, such as agar, to the culture broth. They allow isolated colonies of different bacterial species to be obtained.

In practice, scalar dilutions (in the ratio of 2) of the antibiotic to be tested are added to a series of test tubes containing the culture medium (Mueller–Hinton broth). Each tube is inoculated with a standard amount (105–108 CFU[41]/ml) of the test microorganism.

---

[41]Colony-forming unit

After 18 hours of incubation, the tubes are checked for visible bacterial growth (turbidity): the absence of visible turbidity of the culture medium denotes complete inhibition of microbial growth (see Fig. 5.2).

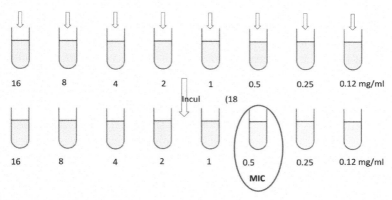

Figure 5.2   Scheme of agar dilution method.

The MBC can be calculated if tubes with no growth are submitted at subculture[42] in fresh medium without antibiotic: the lowest concentration of antibiotic at which the microorganism is unable to grow when transferred to fresh medium equals the MBC.

The agar dilution method is very similar to the culture broth dilution test: plates containing varying amounts of antibiotic are inoculated and growth is assessed.

We can distinguish in macrodilution with a volume of 2 ml or with smaller volumes in microdilution using 96-well micro-titration plate.

The disadvantages mainly concern liquid soils and relate to errors of assessment due to the presence of possible contaminating germs, antimicrobic agent–resistant mutants or non-specific cloudiness of the soil due to substances added to it by the antimicrobial agent.

The method is valid and accurate, but unfortunately also expensive and time-consuming, so its use is limited to a few cases:

---

[42]Subculturing, also referred to as passaging cells, is the removal of the medium and transfer of cells from a previous culture into fresh growth medium, a procedure that enables the further propagation of the cell line or cell strain.

- treatment of very serious diseases in which it is necessary to evaluate the MBC to determine the antibiotic dosage (e.g., in bacterial endocarditis or osteomyelitis)
- evaluation of the susceptibility of slow-growing microorganisms (e.g., mycobacteria and actinomycetes).

There are many accepted guidelines for dilution AST of bacteria, yeast, and fungi.

In 1997, the European Committee on Antimicrobial Susceptibility Testing (EUCAST) unified the different standards previously used in six European countries to interpret the antibiogram. EUCAST is a committee jointly organized by ESCMID (European Society for Clinical Microbiology and Infectious Diseases), ECDC (European Centre for Disease Prevention and Control) and the six formerly active national committees. To date, the clinical breakpoints defined by EUCAST are the only ones to be officially recognized by the EMA (European Medicines Agency), the body that authorizes the marketing of medicines in EU countries.

The reports and documents produced by EUCAST are free of charge and available on the web at http://www.eucast.org.

5.6   Thin-Layer Chromatography: Bioautography

Chromatography is a very effective analytical technique for separating and purifying specific substances in a mixture [11]. Separation is based on chemical and physical interactions at the molecular level between the chromatographic support, the substances to be separated and certain components of the mobile solvent. Separations can take place in a column (chromatographic column) or on a plate of suitable material (TLC).

The applications of chromatography cover a vast field of chemistry, from inorganic to organic, both synthetic and natural. In particular, chromatography is an indispensable technique in synthetic chemistry and natural organic substances, to isolate and purify the components of mixtures of various kinds, and in the biological field to purify polysaccharides, proteins, nucleic acids, viruses, and even cells.

Thin-layer chromatography, or TLC, is a simple chromatographic technique, which makes it particularly suitable for performing qualitative or semi-quantitative analyses. Like all chromatography, it is based on the different separation of different substances between a stationary phase and a mobile phase, depending on the affinity of each substance with them.

All components of the mixture are deposited on the plate (stationary phase) and are transported (eluted) by a moving solvent (mobile phase).

In TLC, the eluent moves from bottom to top by capillarity. The deposited compounds move at different speeds, depending on the different affinity of the analyzed substance toward the eluent and the stationary phase. Separation occurs because the compounds more affinitive to the eluent move more on the plate, while those more affinitive to the stationary phase are more retained (see Fig. 5.3).

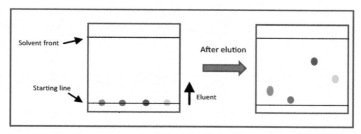

Figure 5.3   Scheme of TLC.

The stationary phase is generally a layer of uniform thickness of about 1 mm of adsorbent material, deposited on a glass or plastic plate. The thickness can be much less if the substrate is aluminum. The adsorbent material can be silica gel, alumina, cellulose powder, or diatomaceous earth powder, all of which are polar materials.

The interaction with the mobile phase depends on the geometry of the constituent particles. We define activity as the ability to adsorb substances in quantitative terms.

The mobile phase is a suitably chosen solvent (or mixture of solvents), which is capable of separating the components of the mixture to be analyzed and is not very similar in polarity to the chosen stationary phase.

The polarity is fundamental for the choice of eluent and it is on this that the extent of substance entrainment along the plate in a TLC depends.

Detection can be carried out in two ways:

- Non-destructive systems: plates with stationary phases with fluorescence indicators sensitive to UV radiation within a certain wavelength range are irradiated with lamps emitting in a band around 254 nm and 366 nm. Observed in the dark, they show a green fluorescence that allows the position of spots or bands (dark in bright field) to be seen. These can be marked with a pencil and the stationary phase can then be removed from the plate by scratching with a spatula.

- Destructive systems: plates are sprayed with solutions of reagents which react with the components to produce colored spots at room temperature or when hot, identifying the different components of the mixture and the standards of comparison. Chemical detectors can be divided into "universal," which are for general use and can detect any compound, and "specific," which react only with certain classes of compounds (e.g., ninhydrin for amino acids, $SbCl_3$ in chloroform for terpenes, steroids, etc.).

The combination of paper chromatography method and contact bioautography was thanks to Goodall and Levi in 1946, to detect different penicillin for their determination [12]. Subsequently, Fischer and Lautner introduced TLC as chromatographic method [13].

This practice combines TLC with both biological and chemical detection methods. This technique is especially used in the screening of organic extracts, mainly plant extracts, for antibacterial and antifungal activity [14].

The TLC plate is sprayed with a microbial suspension. It is then incubated for 48 hours at 25°C under controlled humidity conditions. Tetrazolium salts, which stain in the presence of dehydrogenase living cells, are often used to visualize microbial growth. This is a colorimetric assay that utilizes the ability of mitochondrial dehydrogenases to cleave the tetrazole ring of the yellow MTT (tetrazolium salt) molecule to give a violet formazan salt. The amount of formazan produced is measured in a spectrophotometer and is proportional to the number of living cells [15].

5.7 Antibacterial Activity of Silver Nanoparticles

The first thing to do when you want to develop an antibacterial material is to make sure that there are no cytotoxic effects.[43] Otherwise, we would have no test to verify antibacterial activity.

The most known form of silver toxicity comes from its ingestion or prolonged contact, the disease we are talking about is called argyria [16].

Generally speaking, it is not a life-threatening illness, unless complications occur in internal organs (fortunately rare). However, as argyria gives the skin and mucous membranes an unnatural coloring, it can lead to psychological disorders and can affect the quality of life of the patient.

The exact amount of silver required to cause argyria is not known, and different cases of the disease have been reported over the years, but the manner and timing of exposure to silver or silver compounds has differed from case to case.

Cases of argyria attributable to taking colloidal silver orally— used in these cases as a dietary supplement in various forms of alternative medicine—and nasally (nasal drops) have been reported for periods ranging from 8 months to 5 years.

Colloidal silver is known to have antibacterial, antiviral, and antifungal properties and is therefore used both internally and externally to treat ailments caused by various infections. There are a number of colloidal silver products on the market (food supplements and medical devices), and the use of colloidal silver has also taken off in various alternative medicines. However, in view of the potential side effects that may result from the uncontrolled and uncontrolled use of colloidal silver, it is always advisable to seek the advice of your doctor before using products containing it.

Prolonged and/or excessively high intake of silver compounds causes the metal to be deposited in the skin (particularly in the dermis) and mucous membranes and, following reduction, gives rise to the blue–gray color characteristic of the disease. In addition to the skin and mucous membranes, metal can also be deposited in internal tissues and organs (e.g., spleen, liver, kidneys, central

---

[43]Cytotoxicity: set of biological activities of cells of the immune system leading to death by apoptosis of cells infected with viruses or intracellular or non-host bacteria.

nervous system tissues, bone marrow, etc.), sometimes causing serious symptoms and complications.

The main and sometimes only symptom of argyria is a blue–grayish discoloration of the skin and mucous membranes. This discoloration can be more or less intense and affect one or more parts of the body, as well as manifesting itself in a generalised manner involving the entire body surface.

Argyria usually begins with blue–gray hyperpigmentation of the gums, conjunctival membrane, and nails. The discoloration may then progress, worsening and/or affecting new areas of the body. The course of the disease, however, depends on individual factors and the extent of exposure to silver or its compounds.

The coloring tends to be darker in the areas most exposed to sunlight. To explain this, it has been suggested that silver may somehow stimulate melanocytes to produce more melanin and that this stimulation is increased in the presence of sunlight.

If argyria is present, you should stop using any silver products you are taking. Unfortunately, to date there is no treatment that can cure the disease and eliminate the blue–gray discoloration of the skin and mucous membranes.

In bacterial infections resulting from burns, the gram-negative bacterium *Pseudomonas aeruginosa*[44] is particularly dangerous and resistant to antibiotics. The infection is the main enemy of a wound, as it delays wound healing and promotes chronicity. Due to the considerable invasiveness of some infecting bacterial species, the microbial component may contribute to the worsening of wounds and also of the patient's condition. Several studies have been conducted on the action of silver sulfadiazine against the bacteria present [17].

Although it has been studied for years, it is not yet known exactly how silver performs its antimicrobial action. Explaining the mechanism of action of AgNPs becomes even more complicated.

The bactericidal activity of AgNPs is thought to be due not to a single mechanism of action, but to a combination of several processes, some related to the Ag^+ ions released by the particle, others to the nanoparticle itself.

---

[44]Abbreviated as *P. aeruginosa*

Ag ions bind mainly to thiol groups in the proteins that make up the cell membrane, altering it and consequently compromising the structure of the cell membrane. They can also bind to certain enzymes, thereby neutralizing their action as they no longer have active sites with which to bind, and they can interact with DNA, interfering with correct cell reproduction [18].

The nanoparticle, on the other hand, is capable of attacking and breaking the cell membrane, penetrating it and altering, like Ag ions, the vital functions of the cell [19]. In an interesting work [20] the morphological changes of cells in bacterial cultures of *Escherichia coli*[45] and *Staphylococcus aureus*[46] treated with silver nitrate solutions at various concentrations were studied. TEM[47] images show serious damage to the cytoplasmic membrane, and no cell growth or multiplication was observed after silver ion treatment.

As mentioned above, antibacterial activity is closely linked to the shape and size of the NPs: as the size of the particle decreases, the specific surface area exposed increases and consequently there is a greater release of Ag^+. In addition, the smaller particle size can penetrate the bacteria's cell membrane more easily. AgNPs between 1 nm and 10 nm in size have been shown to have the greatest antibacterial activity [19].

Shape also influences antibacterial activity, with truncated triangle-shaped particles being found to be more effective on *E. coli* than spherical or cylindrical ones [19]. Truncated triangular nanoparticles show bacterial inhibition with silver content of 1 µg. While, in case of spherical nanoparticles total silver content of 12.5 µg is needed. The rod-shaped particles need a total of 50 to 100 µg of silver content. Thus, the AgNPs with different shapes have different effects on bacterial cell.

Another possibility is the generation of ROS (reactive oxygen species) to cause oxidative stress in bacteria. Oxidative stress is a pathological condition that occurs when there is an imbalance between the production and elimination of oxidizing chemical species in a living organism [21].

A free radical is defined as a particularly reactive molecule or atom that contains at least one unpaired electron in its outermost

---

[45]Abbreviated as *E. coli*
[46]Abbreviated as *S. aureus*
[47]Transmission electron microscopy

orbital. Because of this chemical characteristic, free radicals are highly unstable and try to return to equilibrium by stealing from the neighboring atom the electron needed to balance its electromagnetic charge. This mechanism gives rise to new unstable molecules, triggering a chain reaction that, if not stopped in time, will eventually damage cellular structures.

The main reactive oxygen forms (ROS) of biological interest are ozone, superoxide anion, hydrogen peroxide, hydroxyl radical, (alkyl-)peroxyl radical, (alkyl-)hydroperoxide, nitric oxide.

Free radicals react with other physiological molecules, for example, lipids or DNA, in a chain reaction, until two radicals combine to produce a neutral species

AgNPs behave differently depending on whether they interact with gram-positive or negative bacteria.

Gram-positive bacteria have a cell wall with a complex, rigid, and thick peptidoglycan structure. Gram-negative bacteria, on the other hand, have a simpler cell wall structure based on peptidoglycan and also have an outer lipid membrane. We must also distinguish between the various strains of the same bacterium[48] [22].

In studies with gram-negative, *E. coli*, and gram-positive, *S. aureus*, Kim et al. [23] reported greater biocidal efficiency of AgNPs for *E. coli* and ascribed it to difference in cell wall structure between gram-negative and gram-positive microorganisms. However, currently there is insufficient evidence to support such conclusions since most research on bactericidal effect of nanoparticles has been conducted with one or a very limited number of microbial strains.

5.7.1 Test *in vitro* Results

5.7.1.1 Agar disk diffusion test

Initially, these types of tests were carried out. Either by applying the method directly to the discs or by evaluating the inhibition halo by applying it to threads. Figure 5.4 shows photographs of the first tests carried out on viscose and polyester threads treated with the patented method.

---

[48]A bacterial species consists of a group of strains that show a high degree of phenotypic similarity and are distinguished from other groups of related strains by a large number of independent characters.

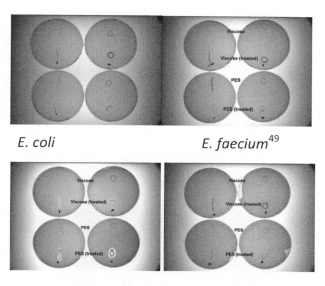

E. coli   E. faecium[49]

P. aeruginosa   S. aureus

Figure 5.4 Agar disk diffusion test for of two materials tested using four bacteria.

Four types of bacteria were tested with two different materials, two gram-positive and two gram-negative, and the MIC values are shown in Table 5.1.

Table 5.1 MIC values of tested samples

Bacterium	Sample size	ZOI of Polyester	ZOI of Polyester-Ag	ZOI of Viscose	ZOI of Viscose-Ag
E. coli	spiral	0	1 mm	0	0
	thread	0	0	0	0
P. aeruginosa	spiral	0	2.7 mm	0	0.2 mm
	thread	0	2 mm	0	0.5 mm
E. faecium	spiral	0	2 mm	0	0.8 mm
	thread	0	2.5 mm	0	1 mm
S. aureus	spiral	0	1 mm	0	0
	thread	0	0	0	0

---

[49]It is the abbreviation of *Enterococcus faecium*.

All samples were analysed in duplicate: 2 cm laid flat on the plate (thread) 6 cm rolled in a spiral (2 complete turns).

From the results, it seems that polyester has a better response to treatment, but this type of test was not very effective due to the size of the yarn to be tested. When testing the treated fabric instead of the yarn, the inhibition zone is much easier to interpret, but the result did not completely satisfy our customers who wanted something more quantitative (see Fig. 5.5).

For this reason, we switched to another type of test.

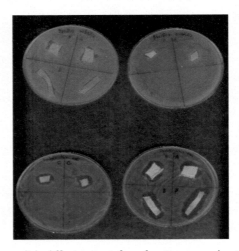

Figure 5.5 Agar disk diffusion test for of cotton samples tested using *B. subtilis*[50] (above) and *E. coli* (bottom).

5.7.1.2   AATCC test method 100–2004

The acronym AATCC stands for American Association of Textile Chemists and Colorists, an association founded in 1921 in Boston by Louis A. Olney, a professor at the Lowell Textile School. This association was created because there was a need for unified testing for the textile industry.

The AATCC 100 test method is designed to quantitatively test the ability of fabrics and textiles to inhibit the growth of microorganisms or kill them, over a 24-hour period of contact. The test microorganism is grown in liquid culture and its concentration is standardized. All

---

[50]It is abbreviation of *Bacillus subtilis*.

the nutritive solution must be sterile. The microbial suspension inoculated on the fabric does not come in contact with solution.

Additional control and test tissues that have been inoculated are left to incubate, undisturbed in sealed jars, for 24 hours. At the end of the incubation period, the microbial concentration can be calculated.

The test can become complicated if the test fabric is hydrophobic.

The antimicrobial activity of the fibers was tested according to officially approved methods with some modifications when necessary. The viable cell count method was employed for quantitative purposes while the zone of inhibition method was employed for qualitative purposes.

Cultures of gram-positive (*S. aureus* and *E. faecium*) and gram-negative (*E. coli* and *P. aeruginosa*) bacteria were grown in Luria Broth for 16–18 h at 37°C to stationary phase, then an aliquot was inoculated in flasks containing fresh Luria Broth to obtain initial optical densities measured at 600 nm (OD600) of 0.05.

The cultures were allowed to grow to reach OD600 values of 0.15 (approx. 30 min) then aliquots of cultures (50 microliters) were absorbed onto the fiber specimens (6 cm). At selected times (i.e., 1, 6, 24 hours) the corresponding specimens were quenched in aqueous NaCl (0.9%, 20 ml). Bacterial growth or killing rates were determined by monitoring the colony forming units (CFU)/ml (which are proportional to OD600 values) in the presence of treated fibers compared to those obtained with untreated fibers (i.e., control).

Quantitatively, the bacteriostatic activity of the treated fibers after a given time (t) was expressed as the percent reduction of bacteria R(t) according to the equation:

$$R(t) = \frac{B(t) - A(t)}{B(t)} * 100 \qquad (5.1)$$

where $A(t)$ was the CFU per ml value observed with treated fibers and $B(t)$ was the CFU/ml value observed in control [24]. Each fibers type was tested in duplicate, and the results were presented as the average of the two tests.

The data obtained with the relative percentages of reduction are shown in Table 5.2. We have indicated with A, B, and C some treated samples.

Table 5.2 Percentage reduction of samples treated with silver nanoparticles

Time (h)	% AATCC standard reduction for *P. aeruginosa*					% Reduction calculated by Jeong for *P. aeruginosa*				
	Sample A	Sample B	Sample C	Sample D	Sample E	Sample A	Sample B	Sample C	Sample D	Sample E
0	-	-	-	-	-	-	-	-	-	-
1	99.84	99.99	99.99	99.39	99.99	99.90	99.99	99.99	99.15	99.99
6	99.99	99.93	98.31	99.59	99.00	99.99	99.99	99.96	99.95	99.97
24	99.91	99.96	99.99	>99.99	>99.99	99.99	99.99	99.99	>99.99	>99.99

Time (h)	% AATCC standard reduction for *S. aureus*					% reduction calculated by Jeong for *S. aureus*				
	Sample A	Sample B	Sample C	Sample D	Sample E	Sample A	Sample B	Sample C	Sample D	Sample E
0	-	-	-	-	-	-	-	-	-	-
1	90.91	84.62	77.14	33.33	50.00	96.30	96.30	91.11	77.78	74.07
6	99.92	99.92	99.92	99.78	94.64	99.99	99.99	99.99	99.99	99.69
24	>99.99	>99.99	>99.99	>99.99	>99.99	>99.99	>99.99	>99.99	>99.99	>99.99

The % standard AATCC reduction is calculated according to the following formula: $100 \, (B - A)/B = \%R$, where A are the cells of the treated sample at relative time, B are the cells at time zero.

The % reduction according to Jeong [24] is calculated as follows: $100 \, (K - A)/K = \%R$, where K are the cells of the untreated sample at relative time.

We can see that the samples tested on *P. aeruginosa* strain have a bactericidal effect, while those tested on *S. aureus* have a bacteriostatic effect.

Figure 5.5 shows the antibacterial activity of AgNPs (size 30 nm) grown on cotton fiber versus obtained by plotting bacterial counts in Log10 CFU/ml as a function of time for *P. aeruginosa*.

Fibers decorated with 30 nm AgNPs exhibit an antimicrobial action related to fiber coverage. If the AgNPs density on the fiber is low ([Ascorbic Acid]/[Ag][51] ratio = 0.7–1.2), a weaker antibacterial activity is observed. Increasing AgNPs density ([Ascorbic Acid]/[Ag] ratio = 2.5–4.5) a full antibacterial activity takes place.

Then we performed a quantitative assay to assess the biocide properties of the deposited films toward gram-positive and gram-negative bacteria using the AATCC 100–1998 method varying the features of the deposited AgNPs. In Figure 5.6, the trend of antibacterial action versus time, plotted using the AATCC100 method are shown. The antibacterial dynamic of the 100 nm AgNPs is different compared with 30 nm AgNPs, probably because of the lower surficial area and coverage. The reduction rate at one hour is lower; the maximum reduction, more than 99% is reached after 6 hours; at 24 hours reduction rate stabilizes around 95%.

For the 30 nm AgNPs reduction rate is influenced by the fiber coverage, high covered fibers show the best reduction rates, higher than 99% after 6 hours.

P. aeruginosa is more susceptible to AgNPs, an antibacterial activity higher than 97% is observed since the first hour regardless the coverage of AgNPs fiber, although highly covered fibers exhibit better biocidal properties. The different efficacy toward gram-positive or gram-negative bacteria is well documented for Ag ions, and generally a larger amount of Ag is required in order to obtain the same efficacy toward gram-positive bacteria [25]. *S. aureus* is a

---

[51]Square brackets indicate the concentration of what is written inside.

typical gram-positive bacterium, with a much thicker peptidoglycan cell wall, compared to *P. aeruginosa* which consequently delays the penetration of Ag ions into the cytoplasm (see Fig. 5.7).

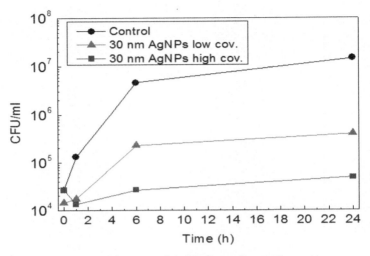

Figure 5.6 Antibacterial activity of AgNPs depending on the coating.

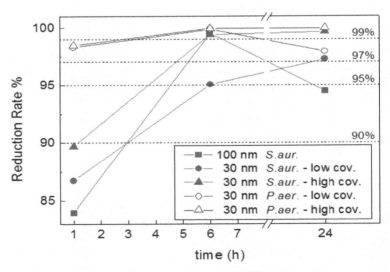

Figure 5.7 Reduction rate of AgNPs versus two different bacteria.

Thanks to ICP–MS analysis (inductively coupled plasma mass spectrometry[52], Thermo Scientific XSeries 2) on the 100 nm AgNPs we checked the amount of Ag on the fibers and the resistance at washing. The amount of Ag respect to total is about 3% (w/w), the reaction yield (Ag$^+$ → AgNPs on the fiber) is about 20%. Amounts of Ag and reaction rates are comparable with the values found by Kelly et al. [26] depending on process parameters. Ag amount on the fibers is much higher compared with other synthesis techniques that range from 0.0029% 40 to 0.014% 41 and 0.57% [27]. Medical Ag product vary from 0.5% in Ag nitrate solution, 1% in Ag sulfadiazine cream and 13% in Acticoat™ [28].

Ranging the number of washing cycles from 0 to 10% Ag on the fiber varies between 2.73% to 0.8% ensuring a high Ag concentration.

In Figure 5.8 is presented a graph showing the relative amount of Ag on the fiber respect to the number of washing cycles. The amount of silver deposited on the fibers was determined also in this case by means of ICP–MS. Quantitative determinations were carried out on fresh samples and on the same samples before and after laundering tests.

Aliquots of each suspension (1 g) were accurately weighed into PTFE flask, 1 ml of concentrated HNO$_3$ was added and the mixture heated to almost dryness. The digested sample was diluted up to 200 mL with 1% (v/v) HNO$_3$. Analysis was performed in duplicate. The results were given as mg Ag / g fiber.

In order to determine the adhesion of AgNPs to fibers, textiles were submitted to laundering tests, that is, 1-cycle, 5-cycles and 10-cycles washing, according to the recommendations of the AATCC [29]. Each washing cycle was performed by using a commercial laundry detergent (5–15% anionic surfactant, less than 5% ionic surfactant) at 20 ml/l concentration, by setting the washing machine to "warm water" (35–40°C) and revolutions to 100 rpm.[53] After each washing cycle the fabrics were rinsed at the same washing temperature for 18 minutes. Finally, the fibers were left to air drying for one day at room temperature.

In Figure 5.8 we can see our results compared with other deposition techniques that exhibit different types of AgNPs fiber bonding: chemically bounded [30] and dipped [31]. Higher the bond

---

[52]Mass spectrometer coupled with a plasma
[53]Revolutions per minute

strength lowers the amount of Ag lost at every washing cycle. AgNPs produced with our method show an adhesion higher than dipped AgNPs but lower than chemically bounded AgNPs.

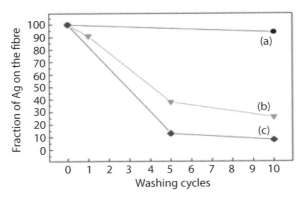

Figure 5.8 (a) Chemical grafted AgNPs [30], (b) Our method, and (c) Dipping of AgNPs [31].

5.8 Small News

In 2015, interesting work was published on the bactericidal action of bacteria killed by ionic silver, bacteria which are therefore capable of killing themselves [32]. In other words, an efficacy far beyond the long-term efficacy that would be desirable for a biocide. For this reason, it has been called the zombie effect.

Bacteria that have been killed by ionic silver show remarkable antimicrobial activity toward the remaining, still viable bacteria. This attitude can be explained in two ways. Firstly, the metal species are not deactivated by the killing mechanism, and so can continue their biocidal effect over and over again; secondly, the dead bacteria are able to release the ionic silver slowly, providing an effective silver reservoir for further action against other living bacteria.

The experiment was conducted as follows: an opportunistic[54] pathogenic bacterium *P. aeruginosa* PAO1 (at a viable bacterial concentration of 108 CFU/ml) was selected and treated with a silver

---

[54]Opportunists, i.e., bacteria that do not trigger pathogenic activity in healthy individuals but are capable of triggering an infection if the individual is immunocompromised.

nitrate solution at a concentration sufficient to kill the bacteria. First by centrifugation and then by filtration, the dead bacteria were separated. At this point a new viable bacterial culture of *P. aeruginosa* of approximately 108 CFU/ml was exposed to the dead bacteria and after 6 hours of exposure, the bacterial viability was counted.

The results obtained are more than satisfactory and could also be achieved with other types of biocides.

The properties of silver as an antiviral are being studied in countries that have experienced the massive spread of viruses. According to that, we believe that the patented treatment may be useful for the containment of airborne infection.

Many scientific papers report on the antiviral properties of silver, and experiments to test them on our materials are underway [33].

References

1. Keyes, K., Lee, M.D., Maurer, J.J. Antibiotics: mode of action, mechanisms of resistance and transfer, In: *Microbial Food Safety in Animal Agriculture: Current Topics* (Torrance ME, Isaacson RE, eds.).

2. Basset, E.J., Keith, M.S., Armelagos, G.J., et al. *Science* (1980) **209**, pp. 1532–1534.

3. Fleming, A. *Br J Exp Pathol* (1929) **10**, pp. 226–236.

4. Saga, T., Yamaguchi, K. *Journal of Japan Medical Association J* (2009) **52**, pp. 103–108.

5. Russell, A.D. *Journal of Antimicrobial Chemotherapy* (1975) **1**, pp. 97–101.

6. https://www.iso.org/home.html

7. Rosselló-Mòra, R., and Amann, R. *Systematic and Applied Microbiology* (2015) **38**, pp. 209–216.

8. Van Teeseling, M.C., de Pedro, M.A., Cava, F. *Frontiers in Microbiology* (2017) **8**, pp. 1264–1283.

9. Austrian, R. *Bacteriological Reviews* (1960) **24** (3), pp. 261–265.

10. Bauer, A.W., Perry, D.M., Kirby W.M.M. *A.M.A. Archives of Internal Medicine* (1959) **104**, pp. 208–216.

11. Snyder, L.R., Kirkland, J.J., Dolan, J.W. *Introduction to Modern Liquid Chromatography*, 3rd Edition (2009), Wiley.

12. Goodall, R., Levi, A. *Nature* (1946) **158**, pp. 675–676.

13. Fischer, R., Lautner, H. *Archiv der Pharmazie* (1961) **294** (1), pp. 1–7.

14. Horváth, G., Jámbor, N., Végh, A., et al. *Flavour and Fragrances Journal* (2010) **25** (3), pp. 178–182.

15. Dewanjee, S., Gangopadhyay, M., Bhattacharya, N., et al. *Journal of Pharmaceutical Analysis* (2015) **5** (2), pp. 75–84.

16. Kwon, H.B., Lee, J.H., Lee, S.H., et al. *Annals of Dermatology* (2009) **21** (3), pp. 308–310.

17. (a) Rosenkranz, H.S., Carranti, S.H. *Antimicrobial agents and Chemotherapy* (1972), pp. 367–372. (b) Carr, H.S., Wlodkowski, T.J., Rosenkranz, H.S. *Antimicrobial agents and Chemotherapy* (1973), pp. 585–587.

18. Rai, M., Yadav, A., Gade, A. *Biotechnology Advances* (2009) **27** (1), pp. 76–83.

19. Pal, S., Tak, Y.K., Song, J.M. *Applied and Environmental Microbiology* (2007) **73** (6), p. 1712.

20. Feng, Q.L., Wu, J., Chen, G.Q., Cui, F.Z., Kim, T.N, Kim J. O. *Journal of Biomedical Materials Research* (2000) **52**, pp. 662–668.

21. Tang, S., Zheng, J. *Advanced Healthcare Materials* (2018) **7**, pp. 1701503–1701513.

22. Rosselló-Mora, R., Amann, R. *FEMS Microbiology Review* (2001) **25** (1), pp. 39–67.

23. Kim, J.S., Kuk, E., Nam, K., et al. *Nanomedicine: Nanotechnology, Biology, and Medicine* (2007) **3**, pp. 95–101.

24. Lee, H.J., Yeo, S.Y., Jeong, S.H. *Journal of Material Science* (2003) **38**, pp. 2199–2204.

25. Kawahara, K., Tsuruda, K., Morishita, M., Uchida, M. *Dental Materials* (2000) **16**, pp. 452–455.

26. Kelly, F.M., Johnston, J.H. *Applied Materials Interfaces* (2011) **3**, pp.1083–1092.

27. El-Shishtawya, R.M., Asiri, A.M., Al-Otaibia, M.M. *Spectrochimica Acta Part A* (2011) **79**, pp.1505–1510.

28. Taylor, P.L., Usshera, A.L., Burrell, R.E. *Biomaterials* (2005) **26**, pp. 7221–7229.

29. AATCC Standardization of Home Laundry Test Conditions AATCC Technical Manual (2007), pp. 407–408.

30. Zhanga, D., Chenb, L.C., Zanga, Y. et al. *Carbohydrate Polymers* (2013) **92**, pp. 2088–2094.

31. Lee, H.J., Yeo, S.Y., Jeong, S.H. *Journal of Materials Science* (2003) **38**, pp. 2199–2204.

32. Ben-Knaz Wakshlak, R., Rami Pedahzur, R., Avnir, D. *Scientific Reports* (2015) **5**, pp. 9555–9560.

33. (a) Galdiero, S., Falanga, A., Vitiello, M., et al. *Molecules* (2011) **16**, pp. 8894–8918. (b) Mori, Y., Ono, T., Miyahira, Y., et al. *Nanoscale Research Letters* (2013) **8**, pp. 93–99. (c) Akbarzadeh, A., Kafshdooz, L., Razban, Z., et al. *Artificial Cells, Nanomedicine, and Biotechnology* (2018) **46**, pp. 263–267. (d) Haggag, E.G., Elshamy, A.M., Rabeh, M.A., et al. *International Journal of Nanomedicine* (2019) **14**, pp. 6217–6229. (e) Xiang, D., Chen, Q., Lin Pang, L., et al. *Journal of Virological Methods* (2011) **178**, pp. 137–142.

Chapter 6

An Exciting and Instructive Experience: Academia–Industry Collaboration

In addition to the already widely discussed antibacterial effect, AgNPs possess other properties of potential interest to an investor.

For example, the utilization of Raman spectroscopy, in particular the SERS effect, could be useful in many areas. Being an inelastic scattering process, the cross section of Raman spectroscopy is particularly low compared to other spectroscopic techniques such as fluorescence spectroscopy or absorption spectroscopy (IR). To overcome this problem, we can use the surface-enhanced Raman scattering spectroscopy, abbreviated as SERS, a form of Raman spectroscopy that overcomes the limitation of low cross sections and increases sensitivity by several orders of magnitude. This spectroscopy technique consists of amplifying the Raman signal of a molecule. The amplification is due to the interaction of the electromagnetic radiation with a metal substrate near which the sample to be analyzed is located [1].

The phenomenon occurs when a metal surface is illuminated by electromagnetic radiation, as a result of which collective oscillations of the conduction electrons with respect to the nuclei of the metal ions are observed on the metal surface: these oscillations are called surface plasmons or localized plasmons. Absorption and scattering can be modified by varying the size and shape of AgNPs.

Silver Nanoparticles: Synthesis, Properties, and Applications
Anna Facibeni
Copyright © 2023 Jenny Stanford Publishing Pte. Ltd.
ISBN 978-981-4968-21-8 (Hardcover), 978-1-003-27895-5 (eBook)
www.jennystanford.com

It would be an interesting thing to use a fiber or a different material on which nanoparticles have grown as a probe in diagnostic medicine too. For example, one could detect differences in signals in the sweat of diabetics [2].

AgNPs can be used to control bacterial growth in a wide range of applications, including burn treatment, dental work, and surgery. They are used in the sterilization of medical consumables such as textiles, food containers, and refrigerators.

The efficiency of AgNPs based fluorescent sensors can be very high and overcome the detection limits. They have been used for the detection of cadmium ions, mercury, some types of antibiotics, nucleic acids [3]. Compared to other noble metal nanoparticles, AgNPs exhibit more advantages for probe, such as higher extinction coefficients, sharper extinction bands, and high field enhancements.

The use of AgNPs is very advantageous if they also have redox catalytic activity in many organic transformations involving the formation of C–C, C–N, C–S, and C–O bonds. They have an interesting catalytic activity in the context of green catalysis [4].

Last but not least can still be used, suitably arranged, as shielding in the case of electromagnetic radiation [5].

Let us now look at some of the industries we have worked with applying our technology.

6.1 Dyeing

The first interest in our patent came from a dyeing company in Como (Italy) specializing in package dyeing of a wide range of fibers: natural, artificial, and synthetic. This company also stands out in the processing of natural yarns with special effects, for example, dyeing with a suede effect on cotton, linen, viscose staple, and wool yarns. The company also has long-standing experience in the dyeing of flame-retardant fibers.

The coil of textile wire has been treated with various concentrations of silver and ascorbic acid. The treated yarns were then examined by SEM and the best ones were subjected to antibacterial tests. The satisfactory results demonstrated the effectiveness of the method at industrial level.

When the silver salt concentration is too high, a kind of nanoparticle crust is observed. Too many nanoparticles do not have a greater antibacterial effect, this is because the nanometric structure is lost and the resulting increased effectiveness (see Fig. 6.1).

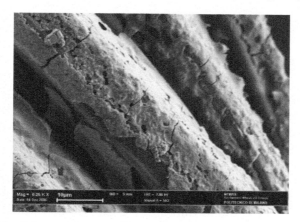

Figure 6.1   Polyester wire coated by too high silver salt concentration.

Another phenomenon was also observed. When the wire is immersed in ascorbic acid solution and then silver nitrate solution is added, the morphology of the nanoparticles changes dramatically, as observed in the literature [6] (see Fig. 6.2).

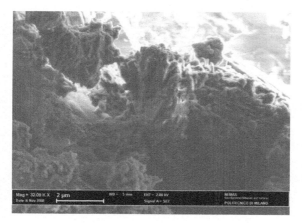

Figure 6.2   Polyester wire coated by a different reaction step.

The advantage for us was that we were able to optimize a process developed in the laboratory that could have posed considerable problems if scaled up on the plant floor.

The partnership with this company unfortunately came to an end, the funding at regional level had ended and there were no economic conditions to continue.

6.2 Mattress Toppers

A mattress topper helps disperse body moisture and prevents allergies by protecting the mattress from dust under the bed. As well as keeping dust and mites out, mattress covers help to combat dirt, one of the main enemies of the mattress, thus making the bed more hygienic. The company involved, in the northeast of Italy, produced mattress covers, on which the spraying process was applied.

Initially, laboratory tests were carried out on polyester, polypropylene and a mixture of polyester and polypropylene. The thicknesses range from 100 g/m^2 to 500 g/m^2. For this reason, the volume of solution they are able to absorb is quite different.

The materials used are all based on hydrophobic polymers, with varying degrees of hydrophobicity, for example, polypropylene cannot absorb the amount of solution sprayed onto polyester. Droplets of liquid are observed on the surface of the former, which change the course of the reaction. Consequently, the two fabrics interact differently with water and must be treated differently.

The reaction parameters that will be changed are the silver salt concentration (nitrate will always be used), the reducing agent concentration, ascorbic acid, the presence, or absence of trisodium citrate (see Fig. 6.3). The spraying distance (~30 cm) and the temperature remain unchanged, no heating is envisaged. The volume of solution will then be adjusted to the materials.

The presence of citrate influences the distribution of nanoparticles for the same concentration of silver and reducing agent. In solution, the citrate has the function of keeping the AgNPs separate from each other. In this case, the reaction volume is much smaller, so the forces involved are different. As a result, the use of citrate plays a key role. At the maximum concentration allowed by the patent (beyond which the disappearance of well-defined AgNPs was observed) the nanoparticles tend to agglomerate into distinct structures. It should be remembered, however, that fiber, even polyester fiber, is not hydrophilic, so areas tend to be created in which the liquid thickens more, creating micro-reactors. The intermediate concentration is the optimum one (see Fig. 6.4).

Figure 6.3 Polyester wires coated by our method: (left) without citrate and (right) with citrate.

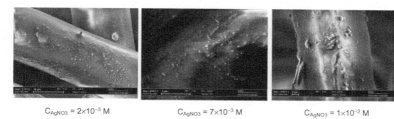

$C_{AgNO3} = 2\times10^{-3}$ M $C_{AgNO3} = 7\times10^{-3}$ M $C_{AgNO3} = 1\times10^{-2}$ M

Figure 6.4 Polyester wires coated by our method at different silver salt concentration: (left) C_{AgNO3} is 2×10^{-3} M, (middle) 7×10^{-3} M, and (right) 1×10^{-1} M.

As reported in the literature, as the concentration of ascorbic acid increases, the structures of AgNPs become more complex (see Fig. 6.5).

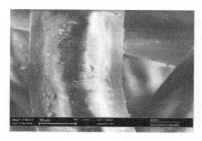 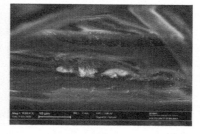

Figure 6.5 Polyester wires coated by our method at different ascorbic acid concentration: (left) $C_{asc.ac.}$ is 4.97×10^{-3} M and (right) 8.47×10^{-2} M.

The material was the so-called non-woven fabric. In practice, this consists of polypropylene microfibers with a diameter of 2 μ. The fabric (polypropylene) runs horizontally to the sprayed side at a speed of 20 m/min, is then passed under a series of IR lamps and comes out almost completely dry. The drying of the material is completed in the air. The presence of silver can already be observed by eye, as there is a color change.

In order to be able to apply the treatment by spraying, a series of nozzles were arranged perpendicular to the flow of the fabric. The main thing was to set up the nozzles at such a distance that a homogeneous coverage of the fabric was achieved. The procedure followed is similar to the previous one: several samples were processed, then analyzed by SEM and the most homogeneous ones were subjected to antibacterial tests. In this case too, industrialization was quite challenging, as the customer explicitly requested that the existing system should not be modified.

The collaboration with the company was very profitable, and the spraying process was developed specifically for use with this. What was required was a spray system that would allow the formation of AgNPs at the speed of the process. The material to be coated runs continuously on a roller. The system on which the test was carried out was not modified at all, only two rows of nozzles were added perpendicular to the surface of the fabric. The sprayers take the different reagents from closed containers placed next to each other.

6.3 Diabetic Socks

Diabetes is a chronic disease that leads to a number of complications, some of which can affect the feet. There may be a risk of developing serious infections in this part of the body. It is well-known that the diabetic foot can lead to the amputation of a toe, foot or even the leg up to the knee, so it is important to take care of your feet. To do this, the choice of diabetic socks is crucial. Diabetic socks have the following characteristics: they must be seamless because seams can cause blisters or ulcers on the skin when rubbing against it. They must absorb moisture, keeping the foot dry is important to prevent infection and fungus. They must be breathable, so made from fabrics that allow moisture to ventilate and keep the foot dry. Hot, diabetes can often cause constriction of the blood vessels, such that poor blood circulation leads to heat loss. Fabrics, which help to keep the foot warm, can improve circulation. Diabetic socks must fit the foot and leg: this prevents the fabric from loosening, bumping into the foot, and causing injury.

The company concerned, also near Como, produced elastic silk socks for diabetics. For this type of pathology, AgNPs could be very interesting. They made it possible to replace the expensive and difficult-to-spin silver wire with a wire coated with nanoparticles.

Silk is an animal textile fiber obtained from cocoons of Lepidoptera: *Heterocera*. *Bombyx mori* is the most important producer of silk cocoons and is reared on a large scale in the domestic state.

Chemically, silk is a protein, fibroin, which is coated on the surface with an irregular layer of sericin, another non-structural protein with the function of a glue, which holds together the fibroin filament, itself composed of an indeterminate number of fibrils. Sericin and fibroin are composed of a series of amino acids (primary structure), which lead to a subsequent secondary structure, either helices or sheets, or strands β.

Fibroin, in its primary structure, consists mainly of the amino acids glycine, alanine, and serine in the form GAGAGS.

Sericin is richer in polar end groups, such as hydroxides, carboxyl, and amine groups, and overall richer in serine, (about 30% of the total amino acids present).

In order to optimize the treatment of silk, certain reaction parameters were varied. The parameters identified have been (a) the concentration of ascorbic acid (reducing agent), (b) the concentration of sodium citrate (stabilizer), and (c) the modulated or unmodulated addition of the reducing agent (slow and fast addiction). Only the silver concentration was kept constant.

The results were more than satisfactory. The ascorbic acid has the function of reducing the silver, the fast addition leads to a more inhomogeneous coating, there seems to be agglomeration of nanoparticles in certain places which then tend to detach. Our method is certainly closer to a slow synthesis, where the silver ions have a chance to reduce and then grow as nanoparticles.

To study the amount of silver grown on the wire, a very sensitive technique is used that can determine various metallic and non-metallic substances present in concentrations of less than one part per billion. It uses an inductively coupled plasma torch to produce ionization and a mass spectrometer to separate and detect the ions produced, abbreviated as ICP–MS.

Table 6.1 shows the silver values found on the wire depending on the variation of certain parameters. The reaction temperature is maintained at 40°C in all cases. The wire was washed in boiling water before treatment, the concentration of ascorbic acid is 9×10^{-3} M.

Table 6.1 Results of ICP–MS analysis on treated wires of different length

Sample	%Ag w/w	Sample weight in g	Ag mg found on the sample	Note
Silk	0.00 ± 0.00	0.0487	0.00	Pristine
S1b40l1	3.10 ± 0.06	0.0480	1.488	Length of 15 m (20 ml of silver nitrate solution)
S1b40l2	1.45 ± 0.02	0.0365	0. 529	Length of 15 m (10 ml of silver nitrate)
S1b40l4	1.12 ± 0.01	0.1365	1.529	Length of 44 m (20 ml of silver nitrate solution)
S1b40l6	1.91 ± 0.01	0.1336	2.552	Length of 30 m (15 ml of silver nitrate solution)
S1	1.77 ± 0.02	0.0842	1.49	Length of 30 m (20 ml of silver nitrate solution)

6.4 Packaging

Now let us move on to another type of industrial sector and material. A company situated near Modena, Italy, called us to apply AgNPs to their packaging, both labels and tissue paper in which the garments are wrapped.

In fashion there are three different types of packaging: customer packaging, logistics packaging, and what we might call accessory packaging.

Customer packaging includes items such as boxes made of paper, cardboard or plastic, shopping bags made of plastic paper or fabric, hangers, often made of plastic, metal or other materials, ribbons, paper tissues, fabric flannels for bags and shoes, and in some cases garment covers. Logistics packaging mainly comprises hangers, cartons, wooden pallets and, particularly important, polybags for clothes covers. What we call accessories are mainly labels, tags and customer gadgets, which strictly speaking do not constitute packaging, but 'travel' with it.

It was a big challenge, the two materials to be processed were very different. Tissue paper is a very light type of paper, which cannot be treated in solution, and required special care. The cotton label was much simpler, but the customer did not want different types of treatment for economic reasons.

Tissue paper feels smooth to the touch without any roughness. The weight (weight to surface ratio) is 21 g/m^2, there are no known surface treatments.

In Fig. 6.6 we can see a sample of tissue paper treated with our method.

Figure 6.6 Tissue paper: (left) not treated and (right) treated with our method.

The printer provided us with two types of ink that are commonly used. One possible strategy was to produce nanoparticles during the molding phase.

The safety data sheets for the products do not give any relevant information on the composition.

The two inks are very different in viscosity, certainly very different from aqueous solutions. The difficulty was in mixing the silver salt and reducing agent solutions with the inks. We can see in Fig. 6.7

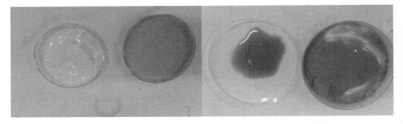

Figure 6.7 (Left) The first ink and (right) the second one. In both figures, the darker ink is the one containing the AgNPs.

The labels were certainly easier to handle, the materials provided were polyester and cotton. In this case we have applied the growth in solution, the fabric does not have the problem of brittleness characteristic of tissue paper (see Fig. 6.8).

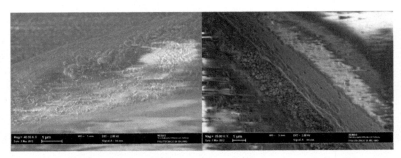

Figure 6.8 Labels treated by our method.

Once again, the application was satisfactory, but the industrialization of the method stops here. Perhaps the packaging market was not yet ready for the exploitation of our application.

6.5 Air Filtration

In this case the company deals with air filtration, in particular using textile pipes. The use of special technical fabrics means that the weight of the micro-perforated pipes is very low, and even in larger diameter applications they rarely exceed 3 kg per linear meter. Such a modest load makes it possible to install these pipes even inside old structures capable of supporting limited overloads, or very light structures such as tensile structures, allowing the installation of efficient distribution systems inside these environments. The problem could be possible moisture in the pipes, which is the ideal environment for bacteria.

The material used in the production of textile pipes is polyester. This type of manufacture is used in civil construction, supermarkets, sports centers, and the food industry. The provided material samples are of two different types, the first has a grammage of 0.0297 g/cm^2, the second one has a grammage of 0.0114 g/cm^2. Both are characterized by grilling and show hydrophobic character.

In this case, two different application techniques were followed: vertical spray and solution synthesis. The solutions, both silver and reducing agent, in both cases have the same starting concentration. Spraying has to be improved, performed with the sample vertically as the hydrophobicity of the material allows only a few nanoparticles to grow. The growth in solution technique is more satisfactory (see Fig. 6.9).

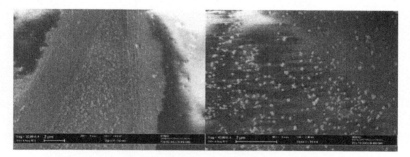

Figure 6.9   Pipes treated by our method in solution.

Silver does not change the chemical structure of polyester, not by binding chemically but by growing on the fiber. Nanoparticles does not affect air filtration, they are very small.

6.6   Wool

In this case, the Biella-based company produces and distributes fine wool, alpaca, cashmere, and silk yarns worldwide. The company spins the material and dyes it. Here, too, the aim was to provide a plus that would distinguish them from their international competitors.

From a chemical point of view, wool fiber is a polymer of protein origin and is made up of keratins, that is, macromolecules resulting from the union of several amino acids linked together by a peptide bond. From a spatial point of view, the keratin chains are arranged in a helix to form a three-dimensional structure.

The surface of wool consists of a layer of 18-methylecosanoic acid, a so-called fatty acid, covalently bonded to the protein layer by means of a thioester bond. Because of this layer, the surface of the wool is hydrophobic, providing a barrier for dyeing and finishing treatments. The flaky structure of wool that we can see in Fig. 6.10, causes resistance to slide against the fiber and gives rise to felt, which is manifested following mechanical stresses exerted when wet: water and bases lift the scales, which thus adhere better to each other, giving rise to the formation of compact felts. Anti-felt treatments involve the deposition of resins on the fiber to level out the scales or chemical oxidants such as hypochlorite, which destroy them.

There are a number of treatments such as plasma and enzymatic treatment to improve the hydrophilic properties of the wool. Acid treatments are also effective.

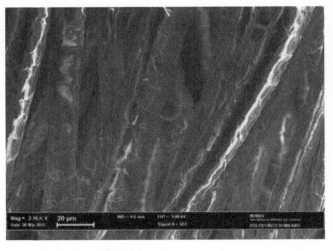

Figure 6.10   SEM image of wool structure.

At pH 3–4 the amino groups are protonated, but not the carboxyl group, which gives the wool a positive charge and increases its hydrophilicity (see Fig. 6.11). Two different types of pre-treatment are tested on traditional wool: basic and acid. The first is used to eliminate the fatty acid that coats the wool, the second to give the wool a positive charge by increasing its hydrophilicity.

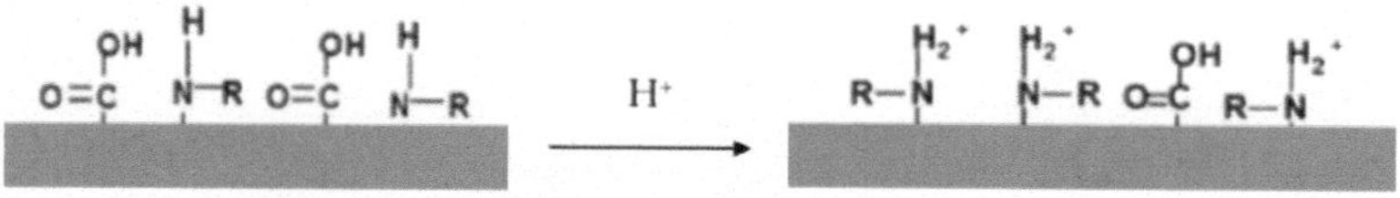

Figure 6.11   Protonation of amino group.

Without pre-treatment, the thread shows an inhomogeneous growth, with the poorly colored thread a sign of a few nanoparticles, the images in Fig. 6.12 show the result.

Figure 6.12   Effect without pre-treatment.

In Fig. 6.13, we can see the result after pre-treatment in a basic environment (KOH 0.1M in methanol). The result is good, the AgNPs grow especially near the scales, the resulting solution is yellowish, and the thread is dark gray.

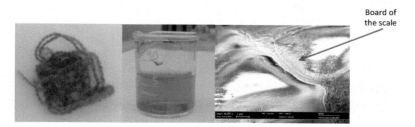

Board of
the scale

Figure 6.13   Effect of basic pre-treatment.

Finally, look at what happens when we carry out a pre-treatment in a 1% aqueous solution of ascorbic acid, which will then be diluted and used as a reducing agent for silver. The resulting solution is grayish, and the wire dark gray (see Fig. 6.14).

Figure 6.14   Effect of acid pre-treatment.

Pre-treatment on raw wool has been shown to enhance the growth of AgNPs, improving the wettability of the material. Pre-treatment with ascorbic acid seems to alter the wool, at least from a macroscopic point of view, no change in the closure of the scales is observed by SEM. The advantage of using ascorbic acid is that it would be sufficient to dilute it and then proceed to reduction.

Precious fibers such as silk and wool (merinos, cashmere) can be used not only for clothing but also by exploiting other properties such as strength, biocompatibility and much. An alternative use for cashmere fabrics can be in the production of accessories (bags, hats, gloves), toys, and in embroidery, for the production of carpets, blankets (especially for children) and upholstery. The coloring imparted by the metallic AgNPs to the material is unavoidable, ranging from yellowish to dark gray through various shades of green and brown, depending on the size and shape of the nanoparticles.

6.7   Electromagnetic Shielding

There are electromagnetic fields in all domestic environments. This is due to the presence of electrical installations, but also and above all to a range of widely used appliances, such as all household appliances, televisions, microwave ovens, radio repeaters, mobile telephones, computers, halogen lamp transformers, hairdryers, and so on. The issue therefore concerns an increasing number of people who are also interested in the exterior of the home, due to the increasing use of mobile phones, electrical, electronic, and telecommunications equipment. The main artificial emissions into the environment are due to radio and television broadcasting and, to a lesser extent, telecommunications equipment.

The question of the possible dangers of non-ionizing electromagnetic fields (EMFs) (i.e., in this between a few hertz and a few hundred gigahertz) emerged after the World War II, as a consequence of the development of applications of this physical agent, initially military environment (radar and telecommunications). Subsequently, the spread of civil EMF applications in industrialized countries (telecommunications, air traffic control, industrial processes, medical diagnosis, and treatment, to name but a few) have resulted in a significant increase in their presence on the territory. This phenomenon has given rise to some concern, even though there are as yet no conclusive data on damage from exposure to electromagnetic fields.

The International Commission on Non-Ionizing Radiation Protection (ICNIRP) is an independent scientific body recognized by the World Health Organisation (WHO) that provides advice and guidance on health protection from exposure to non-ionizing radiation.

ICNIRP publishes guidelines that are the main reference for setting national standards in many countries. In March 2020, it published an update of its guidelines for the protection of the public and workers from exposure to radio-frequency electromagnetic fields. These guidelines apply to many EMF applications, including telecommunication systems including the new 5G technology, for which specific updates have been made for higher frequencies in view of their large-scale use [7].

There are devices based on metals (silver, aluminum, copper) and others on composite materials (polymers and fillers such as carbon-based materials) capable of shielding electromagnetic radiation.

A leading mother and childcare company asked us if it was possible to conduct a study using the shielding action of silver in the form of nanoparticles. By preparing a sufficient coating, discrete shielding values were often obtained. The advantage would be to customise the application directly on the finished product.

In Fig. 6.15, we can see a sample of polyester fabric treated with our method. The size of the tissue is equal to an A4 sheet. On the sample 10 ml of each solution was sprayed.

The study is still in progress.

Figure 6.15   Polyester fabric treated several times with our method.

6.8   More Ideas

The patented method is applicable to many types of material, including polyethylene pellets (see Fig. 6.16) due to the *in situ* growth of AgNPs. As there is no chemical reaction between the silver ion and the treated material, the chemical form of the material is not important.

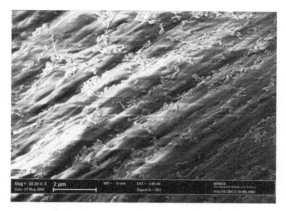

Figure 6.16   Polyethylene pellets treated by our method.

The treatment of the inner shells of motorbike helmets is under development, and here we will have to study the possible release of silver with the sweat always present in this type of activity.

Preliminary studies have also been carried out on commercial mosquito nets, the effect of which could be to contain the spread of malaria (see Fig. 6.17). Mosquito nets are used all over the world; they are the first protection against flies, mosquitoes, and other insects and consequently against possible diseases, especially malaria. The materials used range from polyester to cotton and polyamide. The mesh of the net must be such as to allow air to pass through, to allow good visibility and above all to prevent insects from getting through. Mosquito nets are used as protection on beds, at windows and in camping tents.

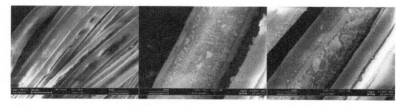

Figure 6.17   Sample of mosquito net: (left) not treated and (middle and right) treated with our method.

Insecticide-treated mosquito nets were developed in the 1980s to prevent malaria. Impregnated with a pyrethroid insecticide such as deltamethrin or permethrin, they kill or repel mosquitoes. They are known by the acronym ITNs (insecticide-treated bed nets). The treatment is recommended for synthetic fabrics periodically needs to be renewed or the mosquito net needs to be replaced, making this an uneconomical solution. There are more durable treatments such as LLINs (long-lasting insecticidal nets) which are more expensive. Mosquito nets should be stored dry and packaged to ensure that the insecticide treatment does not deteriorate. In some countries, mosquito nets are distributed by the government, with instructions for insecticide treatment enclosed.

The treatment can be dangerous if not applied properly, and the disposal of the insecticides also leaves some doubt about safety.

Investigations by energy-dispersive X-ray analysis (EDX) show a high percentage of silver in the material (see Fig. 6.18).

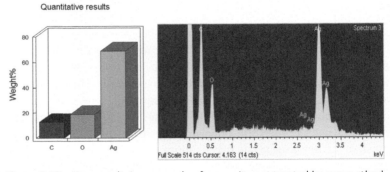

Figure 6.18   X-ray analysis on sample of mosquito net treated by our method.

In a world increasingly at risk from pandemics, using a powerful antibacterial such as silver would certainly help. Treating surgical masks could extend the life of operators and protective equipment.

Spray treatment would allow both the material to be treated before cutting and the masks to be treated individually.

References

1. Campion, A., Kambhampati, P. *Chemical Society Reviews* (1998) **27** (4), pp. 241–250.

2. Moyer, J., Wilson, D., Finkelshtein, I., et al. *Diabetes Technology & Therapeutics* (2012), pp. 398–402.

3. (a) Makwana, B.A., Vyas, D.J., Bhatt, K.D., et al. *Applied Nanoscience* (2016) **6,** pp. 555–566. (b) Sebastian, M., Aravind, A., Mathew, B. *Materials Research Express* (2018) **5,** pp. 085015–085023. (c) Wang, H., Si, X., Wu, T., et al. *Open Chemistry* (2019) **17,** pp. 884–892. (d) Zhang, Y.W., Hai-Long, L., Sun X.P. *Chinese Journal of Analytical Chemistry* (2011) **39** (7), pp. 998–1002.

4. (a) Chang, L.L., Erathodiyil, N., Ying, J.Y. *Accounts of Chemical Research* (2013) **46** (8), pp. 1825–1837. (b) Bhosale, M.A., Bhanage, B.M. *Current Organic Chemistry* (2015) **19** (8), pp. 708–727.

5. Jia, L.C., Yan, D.X., Liu, X., et al. *ACS Applied Materials Interfaces* (2018) **10**, pp. 11941–11949.

6. Fukuyo, T., Imai, H. *Journal of Crystal Growth* (2002) **241**, pp. 193–199.

7. https://www.icnirp.org/

Chapter 7

Conclusions and Perspectives at the End of the Line

The textile supply chain is one of the broadest and most complex, accompanying the product throughout the entire production-distribution process: from the production of the raw material (fiber) to its distribution on the market.

The traditional success factor of Italian textiles is the ability to combine innovation, fashion, and creativity with production technologies.

Industrial textile production requires a wide variety of processes, which differ according to the type of product made, but the starting point is the so-called "textile" fibers, which we can define as "an element characterized by flexibility, fineness, and a high ratio between length and maximum transverse dimension." A common feature of all textile fibers is their "polymeric" chemical structure.

Man-made fibers, produced from petroleum, are suffering the same problems as the petrochemical sector (supply of raw materials, increase in CO_2, global warming, etc.) to these are added concerns about their disposal at the end of the cycle. It is important to respect environmental, social, and economic sustainability.

The textile industry has long attracted public attention for the high impact it has on the environment. The entire supply chain is highly energy intensive and has a substantial impact on CO_2 and

Silver Nanoparticles: Synthesis, Properties, and Applications
Anna Facibeni
Copyright © 2023 Jenny Stanford Publishing Pte. Ltd.
ISBN 978-981-4968-21-8 (Hardcover), 978-1-003-27895-5 (eBook)
www.jennystanford.com

greenhouse gas emissions. Even the production of natural fibers, which are considered more environmentally friendly because they are renewable and easy to dispose of, can have a negative impact on carbon dioxide budgets if cultivation is carried out using intensive, high-input cultivation techniques. Cotton cultivation, for example, is generally intensive, with extensive use of synthetic chemical pesticides, fertilizers, growth stimulants and herbicides. The massive use of these products, often for very long periods due to monoculture regimes, is the direct cause of reduced soil fertility, salinization, loss of biodiversity, water pollution and the emergence of pathogen resistance.

A wide range of chemicals are used in fiber production and subsequent processing. Some of these, such as the pesticides used in cultivation and the dyes used in textile finishing, are harmful to both operators and the environment. When we talk about chemical safety, we mean the requirements that raw materials, production processes and, consequently, finished products must meet in order to guarantee the health of workers and consumers and also the reduction of environmental impact, both in production processes and throughout the life cycle of the products sold.

For these reasons, working in this sector has not been easy, it is clear why it was very important to apply a methodology that did not have too heavy an impact both ecologically and economically.

The use of a reducing agent such as ascorbic acid, which is used in many fields, and a solvent such as water has certainly reduced our impact on the ecosystem. Moreover, in our case, water does not need to be heated because the reduction takes place at room temperature, so thermal pollution is also avoided. The only trick is to use softened water to avoid the precipitation of chlorides, which are poorly soluble salts of silver.

The nanoparticles have helped us enormously, their high reactivity has allowed us to use very low concentrations of silver salt, and here too the environmental and economic savings have been considerable.

The study of how the colloid prepared at higher concentrations interacted with the fiber required many tests. It should be remembered that the classic synthesis was suitably modified so that the nanoparticles would grow on the fiber, not being able to remain in solution.

After the synthesis and characterization, it was time to learn something about bacteria and the tests that are used. The world of biology is really very complex, luckily, I was not alone, an expert was very helpful.

At this point we were ready to try our hand at a patent. Thanks to a study commissioned by the Milan Polytechnic, the writing was not too difficult. The attention that must be paid at this stage concerns the clarity of the patent description, which is very important. You must not omit anything, an expert in the field must be able to reproduce the technique, cover the possible fields of interest and bring out the novelty of your idea.

There may be opposition from the patent applicant, so the method data must be sorted, and there may be a request for new data to support your idea.

Another change was to move from a research-only job to one more focused on contract management, interfacing with customers is often not easy. I had to modify some aspects of the process to meet the needs of the industrial world without distorting it. It was definitely a good exercise to test my mental agility.

My adventure with silver does not end here, over the years other projects have been developed.

We have already talked about the Raman spectroscopic technique used to characterize and identify many materials in a variety of sectors from restoration to pharmaceuticals.

Pellets functionalized with AgNPs could be used as small substrates, which would allow the Raman effect to be maximized even in small containers without contaminating the solution with silver.

Surgical masks (medical devices), those consisting of two or three layers of non-woven fabric (TNT) made of polyester or polypropylene fibers, are normally used in healthcare settings. They are therefore considered disposable, non-reusable, and non-washable. The systems tried so far to sanitize them, in fact, based mainly on the use of dry heat, risk damaging the fabric. These devices offer protection from contamination for a period of 6 to 8 hours of continuous use, after which they must be disposed of.

The masks known as FFP1 (72% filtering capacity), FFP2 (92% filtering capacity), and FFP3 (98% filtering capacity) are personal protective equipment introduced to protect operators from external contamination.

They are available with or without a valve. In the first case, they protect the wearer; in the second, both the wearer and others. These types of masks can be disposable or reusable.

Reusable FFP1, FFP2, and FFP3 masks can be sanitized in two ways, at least as long as the materials they are made of are not subject to wear: by replacing the filters only (according to the manufacturer's instructions on the number of hours the filter provides protection); or by washing them at 60° with a common detergent (according to the manufacturer's instructions, which generally state the maximum number of washes possible so as not to compromise the filtering capacity).

The growth of AgNPs on surgical and non-surgical masks could prolong their life, and tests are in progress to verify that there are no negative effects on perspiration.

Research recently started into the possibility of producing hot water-soluble PVA adhesives containing AgNPs, which would make them antibacterial. In this case, however, the AgNPs were reduced by means of the polyphenols contained in the false fruits of the dog rose (see Fig. 7.1).

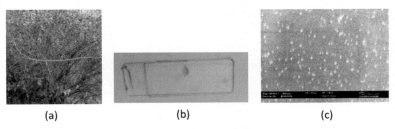

(a)   (b)   (c)

Figure 7.1 (a) Dog rose tree, (b) slides attached with new glue, and (c) SEM image of Ag/PVA glue.

Rosa canina L., known as dog rose, showed the highest antioxidant properties [1]. The presence of polyphenols reduces silver ions to silver nanoparticles, AgNPs [2]. Poly(vinyl alcohol) (PVA) is a polymer widely used, especially in fabric and paper sizing, films for packing, and adhesives. The PVA degradation by gram-negative and positive, is a good thing for biodegradability but it is bad for the injuries, if we use it as adhesive [3].

We have combined the versatility of PVA with the antibacterial action of AgNPs, using a natural reducing agent with anti-inflammatory properties, in particular we have used the fruit of dog rose [4]. The reaction is carried out in an aqueous solution of PVA 3%, in which we have dissolved an appropriate silver nitrate powder, then we added the alcoholic extract of dog rose fruit dried and grounded under agitation.

Our idea is to make a new adhesive, with antibacterial and anti-inflammatory properties, and that can be dissolved by hot water.

This method has been applied also for other molecules such as ZnO, TiO$_2$, and Cu. The results were encouraging, the growth of nanoparticles directly on the fiber could be a good solution in the case of coatings.

References

1. Barros, L., Carvalho, A.M., Morais, J.S., et al. *Food Chemistry* (2010) **120**, pp. 247–254.

2. Careri, M., Mangia, A., Musci, M. *Journal of Chromatography A* (1998) **794**, pp. 263–297.

3. Kawai, F., Hu, X. *Applied Microbiology Biotechnology* (2009) **84**, pp. 227–237.

4. Lattanzio, F., Greco, E., Carretta, D., et al. *Journal of Ethnopharmacology* (2011) **137**, pp. 880–885.

Index